T0141853

Sustainable Ethanol and Climate Change

Muhammad Arshad
Editor

Sustainable Ethanol and Climate Change

Sustainability Assessment for Ethanol Distilleries

Editor
Muhammad Arshad
Jhang-campus
University of Veterinary and Animal Sciences
Lahore, Pakistan

ISBN 978-3-030-59282-0 ISBN 978-3-030-59280-6 (eBook)
https://doi.org/10.1007/978-3-030-59280-6

This Springer imprint is published by the registered company Springer Nature Switzerland AG
The registered company address is: Gewerbestrasse 11, 6330 Cham, Switzerland

The present book is dedicated to

MY FAMILY

Who supported me all the times

Preface

Driven by the desire to secure sustainability and mitigate climate change, bioethanol has been extremely explored in recent decades because it is proved environment friendly. Sustainable and clean fuel is the mandate of Sustainable Development Goals (SDGs), especially the number seven. In pursuit of the SDGs, bioethanol is expected to play a key role, being one of the most sustainable fuel sources.

The major objective of producing this publication is to collate, discuss, and summarize the state-of-the-art knowledge on the current and future prospects of bioethanol's impact on climate change and sustainability. While planning the book, I strived to achieve a comprehensive overview of all aspects of bioethanol from its feedstock, production and usage. It includes the chapters on bioethanol distillery waste (spent wash and CO_2), their disposal, and impact on the environment; and finally a chapter on policy implications. The potential contribution from algae in sequestration of CO_2 released from distilleries and being a sustainable feedstock for bioethanol production has also been discussed.

The knowledgeable and experienced authors for each topic were invited to explain the role of bioethanol to achieve the SDGs. Competent researchers from academia and industry described the role of bioethanol in achieving the SDG-7; the clean and sustainable energy.

I would like to thank all the authors and reviewers for their excellent work and the dedication for keeping the manuscript au fait. The present work is certainly a compliment to numerous authors and they must be proud of such an achievement.

Jhang, Pakistan
June 2020

Muhammad Arshad

Acknowledgements

I would like to send my deepest gratitude to those who have made this book possible, especially the Series Editor for Springer's UN Sustainable Development Goals and all the critiques for their participation in the review of the book proposal and helped me to get this book written in better quality. I would like to express my deep appreciation and indebtedness to all authors for their support in compiling the book.

Also thankful to my family, friends, and others who in one way or another shared their support.

Jhang, Pakistan
June 2020

Muhammad Arshad

Contents

Chapter 1
Sustainable Ethanol Production: An Overview

Muhammad Arshad, Mazhar Abbas, Sadia Javed, Muhammad Adil, Mirza Imran Shahzad, and Munawar Iqbal

Abstract Inevitable depletion in fossil fuel reserves and stringent environmental legislations call for research and development of alternative energy sources. In the past few years, biofuels are emerging as a promising alternative to nonrenewable fossil fuels, thereby causing a minimal impact on the environment. Among biofuels, ethanol has a great potential to come up as an environmentally clean and sustainable transportation fuel to replace petroleum fuel. This book chapter has exhaustively reviewed the literature on the production of bioethanol through various feedstocks ranging from sugary, starchy to cellulosic materials. Various processes used for separation of bioethanol from fermentation broth has been explained. In addition, environmental, societal, and economical aspects of bioethanol production are also discussed. Considering all the facts associated with the latest ethanol production technologies, it is found that still there is a large scope for improvements.

Keywords Bioethanol · *Saccharomyces cerevisiae* · Feedstock · Fermentation · Distillation · SDGs

1.1 Introduction

Energy plays a leading role in the development of social and economic factors (Yildiz et al. 2015). In the present era, the achievements of human beings depend upon the frequent availability of energy resources (Kumar et al. 2015). A significant portion of

M. Arshad (✉) · M. Abbas · M. Adil
Jhang-Campus, University of Veterinary and Animal Sciences Lahore, Lahore, Pakistan
e-mail: muhammad.arshad@uvas.edu.pk

S. Javed
Department of Biochemistry, Government College University, Faisalabad, Pakistan

M. I. Shahzad
Department of Biochemistry and Biotechnology, The Islamia University of Bahawalpur, Bahawalpur, Pakistan

M. Iqbal
Department of Chemistry, The University of Lahore, Lahore, Pakistan

M. Arshad (ed.), *Sustainable Ethanol and Climate Change*,
https://doi.org/10.1007/978-3-030-59280-6_1

the energy is consumed by the transport sector (Saboori et al. 2014). Since hundreds of years people gain energy from gas, petrol, coal, and other oil sources. Currently, these sources share nearly 80% while nuclear sources contribute very low just 6%, and the renewable sources regulate balance (Arshad et al. 2018a). The fossil fuels and industrial gasses pollute the air which, in turn, cause greenhouse effect, and the ambient temperature becomes high which is harmful to animals and plants life (Bano and Arshad 2018). Transportation consumes 27% of energy, which leads to the high consumption of energy among all other consumers (Arshad and Abbas 2018a). According to the World Energy Outlook 2018, by IEA, the transport energy consumption will be about 29% of the primary energy demand in 2030, under the new policies scenario.

The humans themselves are responsible for creating such issues that produce a disturbance in the environment. For the reduction of the role of fossil fuels, the production of renewable sources such like biofuels, photovoltaics, wind and nuclear power is on increase (Twidell and Weir 2015, Arshad and Ahmed 2016, Arshad 2017a). The bioethanol can be used as a substitute for gasoline in the transportation sector as it was the ethanol in an automobile engine as a pioneering fuel (Arshad 2010). A few year bygone, during 1970, when there was an oil crisis, it was considered as potential fuel. The ethanol is very much oxygenated molecule that causes to decline release of dangerous gasses such as carbon monoxide and unburnt organic compounds (Iodice et al. 2016). Due to the utilization of this fuel, the release of these gasses were reduced 10 times in Brazil.

Apart from environmental advantages, the bioethanol has a powerful employment ability. The farming of sugarcane is a source of jobs for the people living in the rural area, especially for poor peoples. The high number of octane and elevated heat of vaporization makes it an attractive transportation fuel. Moreover, its combination along with gasoline elevated the octane number with no aid of any undesirable constituents (Baeyens et al. 2015).

The sugars are the substrate for its production by the anaerobic process. Bioethanol can be produced from the resources composed of the sugars (Arshad et al. 2008). Such supplies can be divided into three basic divisions:

(i) The resources with frequently available sugars include sugar-beet juice, sugarcane juice, and molasses.
(ii) The resources that keep sugars in the form of starches like cassava, cereals, potatoes, and maize.
(iii) The resources with the sugars present in complex form, the cellulosic materials such as wood, rice straw, sugarcane bagasse, and even waste materials.

The production of bioethanol is a very old and well-known process with attractive industrial prominence (Hossain et al. 2017).

This process is shown below.

Saccharomyces cerevisiae is the responsible microbe for the conversion of sugars in into ethanol throughout the world by using various fermentation modes. The *Saccharomyces cerevisiae* can tolerate a high temperature with maximum ethanol level and high tolerances of sugar are useful for ethanol preparation on an industrial scale (Arshad et al. 2011).

The present book chapter is intended to discuss the basic information about ethanol, its production process, feedstock available, industrial-scale processes.

1.2 Feedstock: Accessibility and Limitations

Any biomass containing sugars used for the production of ethanol is called feedstock. Majorly three types of feedstock are categorized as the substrate for ethanol production, sugary feedstocks, starchy feedstocks, and cellulosic feedstocks. Any biomass used as a substrate for the production of ethanol depends upon the fermentation process; pre-fermentation treatment adds additional cost to the overall bioethanol production (Arshad et al. 2018b; Nielsen et al. 2015).

1.2.1 Sugary Feed Stock

Different sugary feedstocks (sugarcane, molasses, beet, and sweet sorghum) were used for the production of ethanol. Sugar in these feedstocks are easy to extract and ferment into ethanol on an industrial scale at an affordable price. Molasses, a byproduct of the sugar industry is one the cheaper source of sugar feedstocks for the production of ethanol (Arshad et al. 2014). As compare to grains as a feedstock, no additional hydrolysis of starch is required. Chemical composition of molasses all over the world shows some variation due to different factor such as soil composition, place where the sugar fields are grown up ambient temperature, season of production, moisture, variety, standard production practices at a particular plant, and by storage variables. Considerable variation in color, viscosity, flavor, and total sugar content of the molasses is also observed.

Besides several advantages, molasses also have some limitations like; High heating of molasses must be avoided since heating over 40.5 °C temperature results in the destruction of sugar, thereby damaging its value. The viscous nature of molasses imparts more difficulties in its handling. Normally molasses is quite safe to handle, but entering of water in the molasses storage tank leads to fermentation. Flammable nature of alcohol vapors is also a great concern. The quality of the molasses is also a serious concern in implementing fuel ethanol policies, since the molasses shows wide variation in terms of its composition, particularly sugar content.

1.2.2 Starchy Feed Stock

Following starchy feedstocks like corn, barley, wheat, rye, rice, potato, millet, triticale, and tubers such as Tapioca (cassava/mandioca) (Adeniyi et al. 2007), were used as biomass for the production of bioethanol. Starch is ahomopolysaccharide, contains glucose as a monosaccharide unit with α (1–4) linkage. Before fermentation (Arshad et al. 2017), starch is treated with dilute H_2SO_4 solely or in combination with enzymes, starch molecules are breakdown into its disaccharides and monosaccharide units (Gulati et al. 1996; Robertson et al. 2006). A major limitation of using the aforementioned feedstock is that it cannot be directly converted into ethanol, first, it should be hydrolyzed by acid or enzymes into fermentable sugar. Capital operating cost increases. Moreover, hydrolysis also takes place at a high-temperature, so energy consumption also increases. In addition to this, the use of corn, wheat, rice, sorghum as a feedstock always raises a debate on food versus fuel.

1.2.3 Cellulosic Feed Stock

Cellulosic biomass is also used as a substrate for the production of bioethanol. Green plant do synthesis up to 50 billion of tons carbohydrates every year at the planet. Claassen et al. (1999) reported that almost 50% of them consist on lignocellulosic material as a biomass, have a remarkable value as a source of feedstock. In cellulosic biomass, a major constituent is cellulose (35–50%), followed by hemicellulose (20–35%) and lignin (12–20%). While small quantities of ash, protein, and lipids are also present (Lynd 1996; Wyman 1999; Greer 2005). On the basis of their complex structure, cellulosic feedstocks are divided into four nonfood categories such as agricultural residues, municipal solid wastes, energy crops, and forestry residues. The use of agricultural feedstock for bioethanol production is encouraged (Arshad et al. 2018c, d).

For the production of bioethanol from renewable energy resources like (cellulosic biomass), pretreatment is required, breakdown of hemicelluloses into simple sugars that can be converted into ethanol. Major advantages of cellulosic feedstock as compared to starch and sugar-based feedstock are cheapest source, easy availability across the world, and the most admirable thing is that they are not used for human food. In addition to this, the biomass like trees and grasses can be grown specifically for ethanol production on marginal lands not suitable for other crops. Moreover, very less fossil fuel sources are required to grow, collect, and convert them into ethanol.

A major limitation in the use of cellulosic feedstock is the presence of lignocellulosic complex structure material. Operating and capital cost also increases due to the breakdown of lignocellulosic material into fermentable sugar. In addition to this, there are challenges in harvesting, collecting, and delivering cellulosic feedstocks.

1.3 Fermentation

Fermentation is an anaerobic biological process in which sugars are converted to alcohol by using microorganisms, preferably yeast. The resulting alcohol from the processes is ethanol. In the distillery, ethanol is produced through the fermentation of molasses. Several microorganisms including *Saccharomyces cerevisiae* (Win et al. 1996; Kiransree et al. 2000; Patrascu et al. 2009), *Kluyveromyces marxianus* (Gough et al. 1998), and *Zymomonas mobilis* (Amutha and Gunasekaran 2001; Panesar et al. 2006; Rebros et al. 2009; Lee and Huang 2000) are available for the production of bioethanol, but the yeast *Saccharomyces cerevisiae* due to its better viability for molasses is employed on a large-scale for the production of bioethanol from cane molasses (Echegaray et al. 2000). Molasses fermentation can be carried out in batch, fed-batch or continuous mode. In the fermentation process, initially, molasses are diluted with water in 1:5 (molasses: water) ratio by volume. The resulting solution is then received in a large tank and the yeast is added to it at 32 °C and incubated for 24–36 h. Fermentation of molasses is a two-step process. In the first step, sucrose present in the molasses is hydrolyzed to glucose and fructose, whereas conversion of glucose and fructose to ethanol and carbon dioxide takes place in the second step. Ruhul et al. (2013) reported that enzymes invertase and zymase present in the yeast *Saccharomyces cerevisiae* acts as a catalyst for the first and second reaction, respectively. These yeasts cannot sustain at a higher alcohol concentration of a product of reaction which consists of ethanol, dissolved solids, suspended solids, unfermentable sugar, sludge, etc. The produced fermented wash is then sent for distillation where the alcohol content of the wash is stripped of and the ethanol of approximately 95.6% (w/w) concentration is recovered.

1.4 Recovery of Ethanol

Ethanol concentration in the fermentation broth is about 5–12% (v/v). Different techniques like distillation and azeotropic distillation is available for the recovery and further concentration of ethanol from the fermentation broth. Such processes minimize the water content in the final ethanol product at lower capital and operating cost as compared to conventional processes. One of the major costs contributing to the production of anhydrous ethanol is the energy cost.

The techniques used for recovery of ethanol from fermentation broth can be broadly classified in three groups as (A) Processes that can be directly employed for the recovery of ethanol from fermentation broth in order to produce commercial ethanol, i.e., Reverse Osmosis. (B) Processes that cannot be employed directly but can be employed in combination for the recovery of ethanol from fermentation broth in order to produce commercial ethanol, i.e., Membrane Separation, and (C) Processes that cannot be employed directly but can be employed particularly after distillation for the recovery of ethanol from fermentation broth in order to produce commercial ethanol. Such processes have their own merits and demerits.

1.4.1 Reverse Osmosis

Reverse Osmosis (RO) is basically a water purification technology that uses a semipermeable membrane in which applied pressure is used to overcome the osmotic pressure. Osmotic pressure is a colligative property driven by a thermodynamic parameter known as chemical potential. In RO, the solute is retained on the pressurized side of the membrane and a pure solvent is allowed to pass through the other side of the membrane. Years back, this technique was used for the separation of ethanol from the fermented wash with the prime objective of replacing the conventional distillation approach in order to make the process less energy-intensive. Since this technology is capable of concentrating the ethanol up to a maximum of 93% (w/w), it cannot be employed in the production of anhydrous ethanol. Additionally, the separation efficiency of RO decreases with an increase in the alcohol concentration in the feed. Various factors affect the separation by RO, however, Choudhury et al. (1985), concluded that the permeability and separation efficiency are the two most important factors governing the performance of the membrane. It is observed that an increase in the operating pressure causes both the flux and separation efficiency to increase. But an increase in the alcohol concentration in a feed causes a considerable drop in permeate flux.

1.4.2 Membrane Separation

In the last few years, PV is used for the separation of ethanol, on the basic principle of solution–diffusion mechanism. PV is a membrane separation process and is considered as one of the most effective and energy-saving process used for the separation of ethanol from azeotropic mixtures. The PV process is classified into vacuum PV and sweep gas PV. The major difference between these categories is that in vacuum PV, partial pressure on downstream (permeate) side of the membrane is reduced by applying vacuum whereas in sweep gas PV, partial pressure on downstream (permeate) side of the membrane is reduced by using an inert sweep gas such as N_2. Membranes can either be hydrophilic or hydrophobic. In general, most membranes are hydrophilic (water permselective) due to water's smaller molecular size, while few membranes are hydrophobic (ethanol permselective). Based on the materials used for membrane production, there are three categories of membranes: (a) inorganic, (b) polymeric, and (c) composite membrane.

1.4.3 *Distillation*

The amount of ethanol in the fermentation medium is about 5–12% (weight/weight). Isolation of ethanol from the fermented broth is largely an energetic procedure. It often utilizes a high amount of the whole energy needs in any distillery. The obtaining of alcohol from fermentation media and it consist of about 95.63% (weight/weight) is recovered isolated by distillation; moreover of ethanol consist of approximately 99.80% (weight/weight) to form fuel alcohol have no chances by purification so the combination of ethanol and water form positive homogeneous azeotrope at this combination (95.63% ethanol and 4.37% H_2O by weight)/ (97.20% ethanol and 2.80% H_2O by volume) at a temperature of 78.15 °C (Huang et al. 2008). The distillation process is very necessary for the production of ethanol, at least 95 GL; centrifuged wine is distilled in the purification and rectification columns, forming hydrous (around 93 weight percent) ethanol. In Brazil, distillation is commonly used for the production of ethanol. Wine is taken at the topmost of the column, present between the two columns. At the top of column D, the components are released by evaporation, on the other hand, vines consist of nearly all water is gained at the lower side of column A. Ethanol-rich streams (phlegm) composed of about fifty weight percent alcohol are gained at columns D and A. The 1st portion of columns is known as distillation. The 2nd portion, also composed of two columns, is named rectification. Phlegm is taken to the columns, and the remaining of H_2O is released at the lower side of column B_1, on the other hand, hydrated alcohol is formed at the topmost of the column.

1.4.3.1 Molecular Sieves Dehydration (MSDH)

Molecular sieve dehydration follows the rules of pressure swing adsorption. The fundamental terminology besides its generation is that electrostatic coordination and polarity are the basic forces among adsorbent and solution. Jeong et al. (2009) suggested that dehydration occurs through the adsorption, ethanol is not utilized but a zeolite bed is liable for discard of H_2O from the liquid alcohol. During this method, liquid alcohol vapors are taken to the zeolite beds; 03 sets are basically utilized, in which one is generating again and the remaining 02 are discarding H_2O, in a cyclic operation. When hydrated ethanol contacts the zeolites, H_2O compounds are adsorbed, on the other hand, anhydrous alcohol is gained at the lower side of the bed. In the reformation phase, water is released from the zeolites by imposing minimum pressure to the bed (Pressure Swing Adsorption), and the gained water–ethanol combination is cycled again to the distillation columns. By using this method more ethanol is formed and also there is lockage of contamination which gives pure product without impurities.

1.5 Sustainability and Socioeconomic Aspects of Ethanol Production

The field of ethanol applications is highly extensive extending from pharmaceutical, potable, industrial to fuel. On average, 73% of global ethanol production relates to fuel ethanol, 17% to beverage alcohol, and 10% for industrial applications. It clears that the main part of the alcohol fermentation is hired as fuel to conserve the conventional fuels.

1.5.1 Sustainable Aspects

Socioeconomic and environmental concerns, heaped on by the accoutrements of global warming with restricted natural resources, have compelled the humans to explore sustainable approaches. In the era of global economic crises, the search for an improved lifestyle has diverted attention toward sustainability. Now sustainability is the demand by all stakeholders of society, including the common man, the government person, and the environmentalists.

All the processes from raw material to production, purchase, sale, marketing, logistics, and consumption must adhere to sustainability. The concept of sustainability is required to be implemented through all walks of life. Global product flow has inserted the concept of sustainability with an emphasis on transport activity.

Therefore, it's need of the day to develop clean liquid transport fuels that can substitute finite fossil fuels to ensure future energy security. The manufacturing of bioethanol requires energy inputs and has an environmental impact. Bioethanol, of the first generation, provide benefits in terms of reducing GHGs and replacing fossil fuel. Certain considerations need to be addressed when assessing the overall efficiency of biofuels, such as competition with food production and release of stored carbon and biodiversity effects as land is cleaned up for increasing energy crops. The combined effects of climate change, persistent instability in fuel prices, the recent food crisis, and global economic deceleration have created a sense of urgency among policymakers, industry, and development practitioners in seeking viable and feasible solutions in the biofuels market.

Biofuel land is used for energy supply. Biofuels compete with other criteria, such as food production, industrial energy, nature conservation, etc., as the bio-productive land area on our planet is small and diminishing. This not only results in higher prices for agricultural and forestry products, but also increases pressure on the environment.

Bioethanol has been as the prime favorable fuel in terms of sustainability over fossil fuels. Its production is associated with diverse socioeconomic benefits and environmental considerations. Bioethanol shows an immense part in reducing the harmful (GHG) productions.

1.5.2 Economic Aspects

To be viable and appreciate commercial approval, the value for alteration of biomass to bioethanol should not be up to existing gas rates, so further consideration should incline to improve the vigor influence of biomass conversion technologies. Pimentel analyzed the rally value of bioethanol from corn in North American nation-state and located out regarding (Echegaray et al. 2000) supplementary energy is required in manufacturing a gallon of alcohol fermentation than energy obtained from a gallon of alcohol fermentation. Pimentel reported that primary studies by the U.S Department of Energy (USDOE) regarding bioethanol manufacturing from biomass described a negative energy output. He further mentioned that these area unit the (GOVT) subsidies that produce the bioethanol reasonable.

However, providing supports for inspiring the association of bioethanol is neither a functional nor a moral process, since the quantity of those subsidies area unit mended by communication an important assessment on tax customers. Eventually these aids area unit liable for increase in different areas of the economy in accumulation, it's established that anhydrous alcohol fermentations are often without delay homogenized with gas; hydrous alcohol fermentation with moderately (200 v/v) of water is not extremely miscible with gasoline. It demands whole dehydration of alcohol fermentation that is not possible by distillation. Purification will distillate alcohol fermentation up (95.6% w/w) exclusively. More dehydration to get rid of remaining (4.5% w/w) needs distinct managements. This added management adds values to bioethanol production. The association value of bioethanol is fairly equalizer by products; like bio-methane, bio-fertilizers just in instance of sauce and dry distillers grains (DDG) in item of corn. However, price exploration of alcohol fermentation from cane sauce while not subsidies, is utterly opposed to that of corn. Production of bioethanol from cane molasses is viable to usual fossil fuels. The arduous assessments are often generated on the evidence of use of feedstock as human food, cropland required; various managements required in its conversion to bioethanol, types of energies desired in its conversion, environmental effluences affected and previous production treatment methods. The inferences concerning to economical advantages of bioethanol manufacturing area unit inadequate or unfair as exclusively some of the aspects that add to total energy control within the alcohol fermentation organization area unit assumed for the valuation. It residences a supplementary strain on the realism that arduous valuation of the bioethanol manufacturing value composed of earlier revealed factors to be done. Value of cropland hired in feedstock farming is also one in all the vital deliberations in bioethanol production. Pimentel revealed that although (100) of the U.S corn is hired for bioethanol production; exclusively 6 June (1944) of recent U.S vehicle fuel claim is achieved through it. Most of the nations within the world area unit abusing to bay kingdoms for oil; visible of being self-employed generous this enslavement, it's essential to assume sure exertions. One in all the constituents of those exertions is to lessen the import of oil leading to the remaining proliferation in fuel alcohol fermentation production. Whatever is added, rise in fuel alcohol fermentation production would cause the rise in infrastructural system within the country.

1.5.3 Social Aspects

Production of bioethanol is associated with different communal impressions starting from well proficient removal of biomass, employability, economic property to develop sector (Chandel et al. 2007). Therefore, all the above influences like economic viability, environmental viability, and employability involving to bioethanol production should be assumed before the operation of fuel alcohol fermentation strategies (Arshad 2017b, Arshad and Abbas 2018b). First and also the biggest advantage of bioethanol production is the well proficient removal of various biomasses which might rather be unused. If not used, the decay of biomass in the atmosphere would cause severe vigor problems. Employability approval is the second vigoros benefit of bioethanol production. Larson et al. reported that bioethanol craft contains a probable to use supplementary entities than normal fossil fuel processing plant. On seeing the whole work series, it's determined that the supporting bioethanol production strategies can be caused for production service occasions in some areas. These areas squeeze grace, assembly, manufacturing, plant operation, tuning, stimulating and fuel proposal. Bioethanol service has its presence, expressively in rural areas of the state. It recommends that it will produce service prospects with supplementary strain in rural areas increasing the economy of those areas by providing careers in fuel yield home-based transference and preservation of development areas.

Rural areas united are those fundamental areas of nation, that part insulant so abundant after in progress and where is a real poverty of progress as there is not any substitute recommend that of producing excluding rural. In rural zones, bioethanol employment offers opportunities not just for literary individuals, however, conjointly for illiterate pioneers together with agrarians and rattan shears. Moreover in the present, accessibility of service opportunities at geographical areas itself stop the passage of the childhoods from rural to urban areas. This saloon of immigration indirectly diminishes the impossible burden on the limited municipal facilities available there. Declaring the remunerations of bioethanol, Demirbas reported that well designed biofuels arise communicate important possessions progress in rural areas.

The biological process yield includes generation of rural occupation, rural electricity deal, loam preservation, etc. Establishment of engagement prospects itself can bring a big economic progress in rural areas. These economic expansions can be finished up in the economically firm generation of rural individuals. In altered confrontations, convenience of service at the agricultural interplanetary prevents the adolescences from earning the money from illegal means, thereby making a peaceful and educated segment in rural areas. Each exertion in that way conjointly makes farming vitality (green energy) be the standard sponsor in economic expansion of some developing countries; as their economy improvement is established on agricultural production and most of their people live inside the rural areas. Implementation of the joined communal development program is so awfully essential. It's strongly believed that combined communal development can push up socioeconomic progress of the country. During this system of these factors energy influence in execution, will definitely bring the prosperity, therefore each exertion has to be produced to support the instinctive content of vigor, service ability and possessions in every time-bound and planned method. During this process, variation of fuel collection might be transporting the money and professions back to the economy.

1.6 Conclusion

Detail review of the bioethanol production technologies reveal that in spite of laboratory-based success stories; high scale production of bioethanol still remains a challenge. The economical competitiveness of fuel ethanol with gasoline is the biggest problem that needs to be tackled today. An economical approach must be there to make the bioethanol without subsidies competitive to gasoline. Research efforts are needed to design and improve the process to produce sustainable and economically feasible fuel to address global fuel needs. Review of the available technologies emphasize that Ethanol could be manufactured from any feedstock containing sugar. The complexity of the conversion processes depends on the structure of sugar and on the nature of the feedstock. Out of the various separation processes, MPRD is found to be the most energy optimized process for the production of extra neutral alcohol (ENA) with steam consumption of 3.0 kg/l of the produced ethanol, whereas hybrid process combining MPRD with MSDH is the best process for fuel (anhydrous) ethanol production. Besides its technical and economical aspects, bioethanol production is associated with various social aspects like rural empowerment, economic sustainability, raised standard of living, prevention of youth migration toward urban areas, etc. Also, the environmental aspects like pollution control, waste disposal, and disposal of biomass are addressed to a better extent. However, in the European continent, where corn is categorized as a major food ingredient; has always resistance for use of corn as a raw material for bioethanol production leading to the continuous debate of food versus fuel. Availability of marginal lands for the growth of energy products in itself is a big concern. Bioethanol production plays an important role in the national interest by providing a reliable alternative to conventional fossil fuels for which most of the countries across the globe have to be dependent on gulf countries. The overall implementation of bioethanol policies will help in improving energy supply security, reducing greenhouse gas emissions, and boosting the rural economies. Moreover, the development of energy crops dedicated to the biofuels production would imply a boost to the agricultural sector (Fig. 1.1).

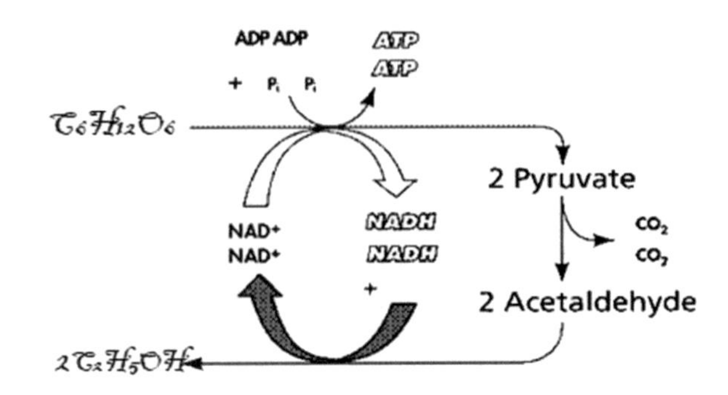

Fig. 1.1 Fermentation of ethanol from glucose

References

Adeniyi OD, Kovo AS, Abdulkareem AS, Chukwudozie C (2007) Ethanol fuel production from cassava as a substitute for gasoline. J Dispersion Sci 28(4):501–504

Amutha R, Gunasekaran P (2001) Production of ethanol from liquefied cassava starch using co-immobilized cells of *Zymomonas mobilis* and *Saccharomyces diastaticus*. J Biosci Bioeng 92(6):560–564

Arshad M (2010) Bioethanol: a sustainable and environment friendly solution for Pakistan. A Sci J. COMSATS–Sci Vision 2011:16–7

Arshad M (2017a) Perspectives on water usage for biofuels production: aquatic contamination and climate change. Springer

Arshad M (2017b) Clean and sustainable energy technologies. In: Clean energy for sustainable development. Elsevier, pp 73–89

Arshad M, Abbas M (2018a) Future biofuel production and water usage. Springer, Cham, Perspectives on water usage for biofuels production, pp 107–121

Arshad M, Abbas M (2018b) Water sustainability issues in biofuel production. Springer, Cham, Perspectives on water usage for biofuels production, pp 55–76

Arshad M, Ahmed S (2016) Cogeneration through bagasse: a renewable strategy to meet the future energy needs. Renew Sustain Energy Rev 54:732–739

Arshad M, Khan Z, Shah F, Rajoka M (2008) Optimization of process variables for minimization of byproduct formation during fermentation of blackstrap molasses to ethanol at industrial scale. Lett Appl Microbiol 47(5):410–414

Arshad M, Zia MA, Asghar M, Bhatti H (2011) Improving bio-ethanol yield: using virginiamycin and sodium flouride at a Pakistani distillery. Afr J Biotech 10(53):11071

Arshad M, Ahmed S, Zia MA, Rajoka MI (2014) Kinetics and thermodynamics of ethanol production by Saccharomyces cerevisiae MLD10 using molasses. Appl Biochem Biotechnol 172(5):2455–2464

Arshad M, Hussain T, Iqbal M, Abbas M (2017) Enhanced ethanol production at commercial scale from molasses using high gravity technology by mutant *S. cerevisiae*. Braz J Microbiol 48(3):403–439

Arshad M, Bano I, Khan N, Shahzad MI, Younus M, Abbas M (2018a) Electricity generation from biogas of poultry waste: An assessment of potential and feasibility in Pakistan. Renew Sustain Energy Rev 81:1241–1246

Arshad M, Bano I, Younus M, Khan A, Rahman A (2018b) Health concerns associated with biofuel production. Springer, Cham, Perspectives on water usage for biofuels production, pp 97–105

Arshad M, Zia MA, Shah FA, Ahmad M (2018c) An overview of biofuel. Springer, Perspectives on water usage for biofuels production, pp 1–37

Arshad M, Bano I, Younus M, Khan A, Rahman A (2018d) Health concerns associated with biofuel production. Perspectives on water usage for biofuels production.Springer, Cham, pp 97–105

Baeyens J, Kang Q, Appels L, Dewil R, Lv Y, Tan T (2015) Challenges and opportunities in improving the production of bio-ethanol. Prog Energy Combust Sci 47:60–88
Bano I, Arshad M (2018) Climatic changes impact on water availability. Springer, Cham, Perspectives on water usage for biofuels production, pp 39–54
Chandel AK, Chan ES, Rudravaram R, Narasu ML, Rao LV, Ravindra P (2007) Economics and environmental impact of bioethanol production technologies: an appraisal. Biotechnol Mol Biol Rev 2(1):14–32
Choudhury JP, Ghosh P, Guha BK (1985) Separation of ethanol from ethanol—water mixture by reverse osmosis. Biotechnol Bioeng 27(7):1081–1084
Claassen PAM, Van-Lier JB, Contreras AL, Van-Niel EWJ, Sijtsma L, Stams AJM, De- Vries SS, Weusthuis RA (1999) Utilisation of biomass for the supply of energy carriers. Appl Microbiol Biotechnol 52(6):741–755
Echegaray OF, Carvalho JCM, Fernandes ANR, Sato S, Aquarone E, Vitolo M (2000) Fed-atch culture of *Sacchoromyces cerevisiae* in sugar-cane blackstrap molasses: invertase activity of intact cells in ethanol fermentation. Biomass Bioenerg 19(1):39–50
Gough S, Barron N, Zubov AL, Lozinsky VI, McHale AP (1998) Production of ethanol from molasses at 45 °C using Kluyveromyces marxianus IMB3 immobilized in calcium alginate gels and poly (vinyl alcohol) cryogel. Bioprocess Eng 19(2):87–90
Greer D (2005) Creating cellulosic ethanol: spinning straw into fuel. Biocycle 46(4):61–65
Gulati M, Kohlmann K, Ladisch MR, Hespell R, Bothast RJ (1996) Assessment of ethanol production options for corn products. Biores Technol 58(3):253–264
Hossain N, Zaini JH, Mahlia TM (2017) A review of bioethanol production from plant based waste biomass by yeast fermentation. Int J Technol 8(1):5–18
Huang HJ, Ramaswamy S, Tschirner UW, Ramarao BV (2008) A review of separation technologies in current and future biorefineries. Sep Purif Technol 62(1):1–21
Iodice P, Senatore A, Langella G, Amoresano A (2016) Effect of ethanol–gasoline blends on CO and HC emissions in last generation SI engines within the cold-start transient: an experimental investigation. Appl Energy 179:182–190
Jeong JS, Jang BU, Kim YR, Chung BW, Choi GW (2009) Production of dehydrated fuel ethanol by pressure swing adsorption process in the pilot plant. Korean J Chem Eng 26(5):1308–1312
Kiransree N, Sridhar M, Rao LV (2000) Characterization of thermo tolerant, ethanol tolerant fermentative Saccharomyces cerevisiae for ethanol production. Bioprocess Eng 22(3):243–246
Kumar A, Kumar N, Baredar P, Shukla A (2015) A review on biomass energy resources, potential, conversion and policy in India. Renew Sustain Energy Rev 45:530–539
Lee WC, Huang CT (2000) Modelling of ethanol fermentation using *Zymomonas mobilis* ATCC 10988 grown on the media containing glucose and fructose. Biochem Eng J 4(3):217–227
Lynd LR (1996) Overview and evaluation of fuel ethanol from cellulosic biomass: technology, economics, the environment, and policy. Annu Rev Energy Env 21(1):403–465
Nielsen F, Tomas-Pejo E, Olsson L, Wallberg O (2015) Short-term adaptation during propagation improves the performance of xylose-fermenting *Saccharomyces cerevisiae* in simultaneous saccharification and co-fermentation. Biotechnol Biofuels 8(1):219
Panesar PS, Marwaha SS, Kennedy JF (2006) Zymomonas mobilis: an alternative ethanol producer. J Chem Technol Biotechnol 81(4):623–635
Patrascu E, Rapeanu G, Bonciu C, Vicol C, Bahrim G (2009) Investigation of yeast performances in the fermentation of beet and cane molasses to ethanol production. Ovidius Univ Ann Chem 20(2):199–204
Rebros M, Rosenberg M, Grosova Z, Paluch M, Sipocz M (2009) Ethanol production from starch hydrolyzates using *Zymomonas mobilis* and glucoamylase entrapped in polyvinyl alcohol hydrogel. Appl Biochem Biotechnol 158(3):561–570
Robertson GH, Wong DW, Lee CC, Wagschal K, Smith MR, Orts WJ (2006) Native or raw starch digestion: a key step in energy efficient biorefining of grain. J Agric Food Chem 54(2):353–365
Ruhul AM, Saquib HM, Sarker M (2013) Simulation of ethanol production by fermentation of molasses. J Eng 1(4):69–73

Saboori B, Sapri M, bin Baba M (2014) Economic growth, energy consumption and CO_2 emissions in OECD (Organization for Economic Co-operation and Development's) transport sector: a fully modified bi-directional relationship approach. Energy 66:150–161

Twidell J, Weir T (2015) Renewable energy resources. Routledge

Win SS, Impoolsup A, Noomhorm A (1996) Growth kinetics of *saccharomyces cerevisiae* in batch and fedbatch cultivation using sugarcane molasses and glucose syrup from cassava starch. J Ind Microbiol 16(2):117–123

Wyman CE (1999) Biomass ethanol: technical progress, opportunities, and commercial challenges. Annu Rev Energy Env 24(1):189–226

Yildiz O, Rommel J, Debor S, Holstenkamp L, Mey F, Muller JR, Radtke J, Rognli J (2015) Renewable energy cooperatives as gatekeepers or facilitators? Recent developments in Germany and a multidisciplinary research agenda. Energy Res Soc Sci 6:59–73

Chapter 2
Sustainable Approaches Toward the Production of Bioethanol from Biomass

Muhammad Faisal and Aamer Saeed

Abstract In the coming future, natural energy sources such as diesel and petrol are going to be diminished. Therefore, interest in alternate sources of reliable, affordable, and sustainable energy has prompted much attention in the past three decades. Bioethanol is one of the promising alternatives, which is obtained from different biomass feedstocks. Bioethanol production from biomass is a well-established technology. This chapter provides a detailed insight on the current technologies for sustainable production of bioethanol from Indian bamboo using enzymatic saccharification technique, sweet sorghum bagasse through microwave irradiation approach, oil palm frond through ultrasonic irradiation method, sweet sorghum stalk via advanced solid-state fermentation system (ASSF), lignocellulosic through low-intensity pulsed ultrasonic methodology, sugarcane bagasse through hydrolysis approach using yeast cellulolytic enzymes, water hyacinth *Eichhornia crassipes* using an approach based on the preparation of the saccharified solution, sugarcane bagasse through sono-mediated enzymatic saccharification protocol, reed straw and corn stover using liquid hot water (LHW) pretreatment method, wheat straw using gamma irradiation approach, sugarcane bagasse through consolidated bioprocessing technique using *Phlebia* sp. MG-60, waste of date palm fruit using solar energy, and mandarin peel wastes using steam explosion method. Further, the chapter discusses industrial scale formation processes of bioethanol from sugar-, lignocellulosic-, and starch-mediated feedstocks. Moreover, the chapter also summarizes the commonly used feedstocks for sustainable and cost-effective bioethanol production and the biochemical reactions involved in the transformation of vegetative biomass to bioethanol. Pros and Cons of each sustainable approach for bioethanol production are discussed in detail, which will be very fruitful for the research community to get knowledge about the finest technologies, improve traditional methods, and develop further new more reliable, affordable, and sustainable methodologies for bioethanol production.

Keywords Sustainable production · Bioethanol · Biomass · Sugar · Lignocellulosic · Starch

M. Faisal · A. Saeed (✉)
Department of Chemistry, Quaid-i-Azam University-45320, Islamabad, Pakistan
e-mail: aamersaeed@yahoo.com

M. Arshad (ed.), *Sustainable Ethanol and Climate Change*,
https://doi.org/10.1007/978-3-030-59280-6_2

- Bioethanol is one of the promising energy alternatives, which is obtained from different biomasses and implementation of modern technologies to produce bioethanol can have a large influence on the energy system.
- Over the past few decades, researches are developing highly effective technologies to advance the process of bioethanol production from different biomass in order to implement SDG-7.
- Formation of bioethanol from different biomass including Indian bamboo, sweet sorghum bagasse, oil palm, sweet sorghum stalk, water hyacinth, lemon peel, sugarcane bagasse, corn stover, reed straw, and wheat straw.
- Application of modern technologies for bioethanol production such as sono-mediated enzymatic saccharification, microwave irradiation, ultrasonic irradiation, steam explosion, and gamma irradiation.
- An industrial scale formation processes of bioethanol production from sugar-, lignocellulosic-, and starch-mediated feedstocks.
- Commonly used feedstocks for sustainable and cost-effective bioethanol production.

2.1 Introduction

Scientific developments have offered mankind with different ways to use resources to advance the quality of life. A development is "sustainable" if it "fulfills the demands of the present energy without compromising the ability of the forthcoming generations to satisfy their own needs" (Brundtland 1987). Preference of unsustainable alternatives, along with the ever-increasing population of the world has led to the depletion of resources. In 2015, the population of the world reached up to 7.3 billion and expected to increase by 33% to cross up to 9.7 billion in 2050, and by 53% to reach up to 11.2 billion in 2100 (Melorose et al. 2015). The energy demand fulfillment of growing population is one of the great concerns of current world and a big challenge (Pacala and Socolow 2004). Burning of hydrocarbons, viz., natural gas, coal, and oil is the source of most of the global energy. In 2015, these fuels (hydrocarbons) accounted for 86% of the energy consumed (Melorose et al. 2015; Akorede et al. 2010). Natural gas, coal, and oil are sustainable and finite energy sources and their global supplies are unlikely to last more than 120 years (Shafiee and Topal 2009), if consumed at the current rate of consumption. Furthermore, consumption of fossil fuels is one of the most important causes of greenhouse gasses (GHG), whose destructive effects on global warming and climate change are well reported and forecasted. In an attempt to decrease GHG emissions, and to supplement fossil fuels, clean and low-carbon alternatives are being advanced and deployed. The rising energy demand, insecurity of fossil fuel supply, together with favorable policy preferences, federal mandates, and associated subsidies (Guo et al. 2015; Pérez-Lombard et al. 2008), remain key drivers in the diversification of energy sources and gradual shift in balance toward renewable energy. These include energy from wind, geothermal, hydrothermal, nuclear, solar, and biomass, which have grown suddenly over the past 15 years (Guo et al. 2015; Akorede et al. 2010), and continue to appeal to investors, facilitating rapid developments in terms of infrastructure and technology.

The energy obtained via the transformation of renewable organic materials from plant or animal sources ("biomass") to energy or energy-containing compounds is known as "bioenergy". Modern bioenergy utilizes extremely sustainable and efficient conversion processes as compared to the low effective classical bioenergy that contains combustion of unprocessed manures straws or wood (Gurung and Oh 2013). In 2014, the global total primary energy supply (TPES) was 5.7×1020 J, out of which bioenergy accounted for 10%, making it the largest renewable energy source (Guo et al. 2015; Pérez-Lombard et al. 2008). The particular interest drawn toward bioenergy can be owing to three reasons, i.e., (i) the richness of biomass (global bioenergy potential of land excluding agriculture lands, wilderness, infrastructure, and forests is sufficient to fulfill 1/3 of the present global energy demand), (ii) convenience of well-known infrastructure and processes, and (iii) biomass remaining the sole feedstock for the large-scale liquid biofuels production (Guo et al. 2015). Global bioenergy consumption will continue to increase in the following decades and is calculated to supply as much as 30% of global energy in 2050 (Guo et al. 2015). Sustainable processing of biomass to afford bioenergy, biochemical, and biofuels is referred to as biorefining (Mielenz 2001). Biorefineries can be regarded as environment-friendly analogs of oil refineries. They are constructed to diminish waste via effective consumption of every fraction of the biomass, with separate conversion pathways to process each fraction (Menon and Rao 2012). The value-added bioproducts serve to offset biofuel production costs and recover profitability of the industry. By 2020, biorefineries are calculated to produce a global revenue of US$295 billion across the biomass value chain (Bozell 2008). Biofuels can be defined as gaseous or liquid fuels derived from biomass. Liquid biofuels include bioethanol, biopropanol, biobutanol, and biodiesel, while biohydrogen and biomethane are common gaseous biofuels. Production of biofuels is expected to amass US$80 billion in revenue by 2020 (Bozell 2008), and its sustainable production can reduce carbon dioxide emissions by 2.1 Gt annually (Tanaka 2011). The contribution is expected to be critical, especially in transportation, fueling 27% of the sector's demand by 2050 (Thiruvengadathan 2017).

Bioethanol is known as the most broadly used biofuel in the transportation sector and has a long history as alternative fuels. In 1984, France and Germany started to use bioethanol as a fuel in internal combustion engines (ICEs) (Demirbas and Karslioglu 2007). Utilization of bioethanol by Brazil was initiated since 1925. In the United States and Europe, bioethanol was extensively employed until the early 1900s. After World War II, the application of bioethanol was neglected due to its expensive manufacturing cost as compared to petroleum fuel until the oil crisis in 1970 (Demirbas 2009). The attention in employing bioethanol has been growing since the 1980s, and in many countries, it has been regarded as an alternative fuel. Global bioethanol manufacturing raised from 13.12 billions of gallons in 2006 to almost 25.68 billions of gallons in 2015, with a slight decrease in 2012 and 2013 (Thiruvengadathan 2017; Menon and Rao 2012; Balat and Balat 2009). The USA is the largest bioethanol manufacturer with the manufacturing of approximately 15 billion gallons in 2015.

Bioethanol is also called as ethyl alcohol or chemically C_2H_5OH or EtOH. It can be employed directly as pure bioethanol or mixed with the gasoline to produce "gasohol" (Staniszewski et al. 2007). It can be employed as an octane enhancer or gasoline improver and in bioethanol–diesel blends to decrease the emission of exhaust gasses (Pejin et al. 2009). Bioethanol offers numerous benefits over gasoline, for instance, higher octane number (viz. 108), higher flame speeds, broader flammability limits, and increased heats of vaporization (Balat and Balat 2009). In comparison with petroleum fuel, bioethanol is readily biodegradable, less toxic, and produces lesser air-borne pollutants (John et al. 2011). A variety of biomass feedstocks from the first, second, and third-generation has been used in bioethanol production (Moriarty and Honnery 2017). The first-generation bioethanol involves feedstocks rich in sucrose (sweet sorghum, sugar beet, sugarcane, and fruits) and starch (barley, sweet potato, potato, cassava, rice, corn, and wheat). Second-generation bioethanol comes from lignocellulosic biomass, for example, straw, wood, and grasses. Third-generation bioethanol has been derived from algal biomass including microalgae and macroalgae (Fig. 2.1) (Nigam and Singh 2011). Among all the production approaches in the first, second, and third generation, the sustainable approaches are highly valuable (Smith and Porter 2018). This chapter covers the summary of all sustainable approaches which are applied to produce bioethanol. Pros and Cons of each technique and explanatory material of scope of each approach will be very fruitful for the research community to increase the diversity of procedures for bioethanol production, to get knowledge about the most optimum and recent approaches toward bioethanol production, to select a finest synthetic approach for bioethanol production, to improve traditional synthetic approaches for bioethanol production, and to develop further new methodologies for bioethanol production.

2.2 Commonly Used Feedstocks for Sustainable Bioethanol Production

For centuries, bioethanol has been employed as a source of energy. Bioethanol can be obtained from different kinds of feedstock. Frequently employed feedstocks for the production of bioethanol across the world can be produced from sugar-mediated feedstocks (for example sugarcane, sweet sorghum, and sugar beet) that are rich in fermentable sugars, or feedstocks which are rich in polysaccharides sugars that are subsequently hydrolyzed to supply the fermentable sugars for the production of bioethanol (Thiruvengadathan 2017). The latter can be starch-mediated feedstocks (for example wheat, corn and rice), or feedstocks containing a complex of cellulose, hemicellulose, and/or other polysaccharides (lignocellulosic biomass). The cost of bioethanol production processes mostly relies on the kind of feedstock being employed as the raw material. Lignocellulosic biomass, which contains numerous polysaccharides, can also be employed to produce bioethanol. The following section describes the frequently employed feedstock, which are used to produce bioethanol.

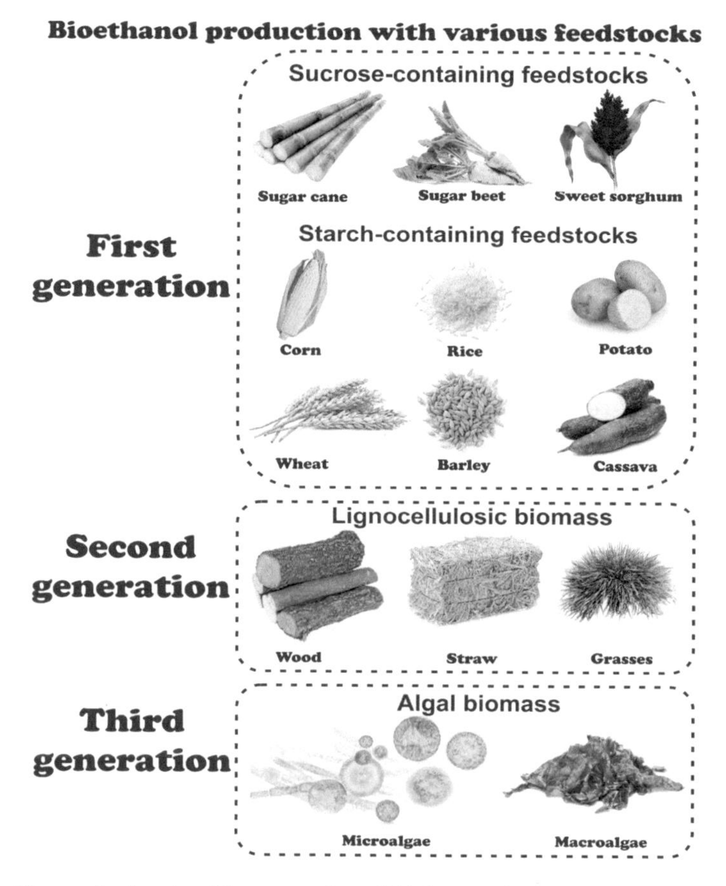

Fig. 2.1 Biomass feedstocks of first-, second-, and third-generation

2.2.1 Sugar or Sucrose-Mediated Feedstocks

Most of the world's bioethanol is obtained from sugar crops. The transformation process of sucrose to bioethanol is more direct than starch to bioethanol. The disaccharide sucrose can be hydrolyzed using invertase enzymes secreted by yeasts to afford the readily fermentable sugars, fructose and glucose. The hydrolysis can also be partially attained during the juice conditioning step. Commercial sucrose-mediated energy crops are sugarcane, sweet sorghum, and sugar beet (Cardona and Sánchez 2007; Zabed et al. 2014).

Sugarcane (*Saccharum officinarum*) is a C4 crop (capable of highly effective carbon fixation that avoids photorespiration, with four carbon acids formed as its first product) and is the most significant feedstock used for the production of bioethanol in subtropical and tropical countries (Table 2.1). It is employed for bioethanol production either as sugarcane juice (as in Brazil) or as molasses (as in India), a noncrystalline by-product of sucrose purification. The sugarcane juice has a fermentable

Table 2.1 Potential bioethanol yields from different feedstocks (Balat et al. 2008; Guo et al. 2015)

Entry	Feedstock	Bioethanol yield potential L/ton	Entry	Feedstock	Bioethanol yield potential L/ton
1	Sugarcane	70	6	Cassava	180
2	Sugar beet	110	7	Corn	360
3	Sweet sorghum	60	8	Rice	430
4	Sweet potato	125	9	Barley	250
5	Potato	110	10	Wheat	340

sugar content between 12 and 17%, more than 90% of which is composed of sucrose. The juice also contains organic nutrients and minerals in minute amounts which are conducive for microbial growth. The sugarcane mediated Brazilian bioethanol industry represents a robust renewable energy model, consistently achieving at least twice the renewable energy produced per fossil fuel consumed (RER) ratio than the maximum RER achieved by any US bioethanol plant based on corn (Cardona and Sánchez 2007; Astolfi-Filho et al. 2011; Chum et al. 2014; Ergun and Mutlu 2000; Ghosh and Ghose 2003; Laluce et al. 2016).

Sugar beet (*Beta vulgaris*) is a crop which grows in temperate climatic environments, most common in European countries. It requires 35–40% lesser water and fertilizer than sugarcane. The raw juice extracted from sugar beet, as well as its by-product from the sugar industry, beet molasses, are sources of fermentable sugars for bioethanol production. Sugar beet juice contains 16.5% sucrose (Ogbonna et al. 2001), and as around 85–90% of its sugars are fermentable, the juice can be employed directly after pH adjustment, making it a convenient substrate (Balat et al. 2008; Dodić et al. 2012; Ergun and Mutlu 2000).

Sweet sorghum (*Sorghum bicolor* L.) is a C4 crop, capable of growing in both temperate and tropical climatic environments (Table 2.1). The crop's grain, as well as stalk juice, can be employed for the production of bioethanol. Sweet sorghum juice has a fermentable sugar content of around 13 to 17% sugars, 10 to 14% of which consists of sucrose (Akbulut and Özcan 2008), with reducing sugars predominantly contributing the rest. The juice also comprises of micronutrients that can increase yeast growth and metabolism (Cao et al. 2006). Sweet sorghum possesses numerous benefits when compared to other biofuel feedstocks, for instance, high carbon assimilation, lower nitrogen and fertilizer requirements, high tolerance to drought and cold temperatures, high photosynthetic efficiency, and short growth cycles (3.5 months) (Kim and Donal 2011). Furthermore, it has the highest stalk juice extractability (71.9%) among all sugar crops, and the crop can potentially achieve a bioethanol yield of up to 8000 L ha^{-1}, which is twice that of corn and 30% higher than sugarcane (Deesuth et al. 2012), making it a promising feedstock for obtaining bioethanol (Thiruvengadathan 2017; Andrzejewski et al. 2013; Barcelos et al. 2016; Kumar et al. 2013; Laopaiboon et al. 2009).

2.2.2 Starch-Mediated Feedstocks

Starch is a long chain homo-polymer of D-glucose, which is hydrolyzed to attain glucose syrup suitable for bioethanol production, a process frequent in Europe and North America (Balat et al. 2008). Wheat and corn and are the principal starch-mediated feedstocks. Other starchy feedstocks employed for bioethanol production include rice, cassava, potato, sweet potato, and barely (Table 2.1).

Corn (*Zea mays L.*) is a Mexican-native giant C4 grass, and the most grown grain in the Americas (Matsuoka et al. 2002). The stalk of the plant comprises of sugars, which on maturity accumulate in the corn kernel as starch (Coelho and Dale 1980). The starchy kernel is either processed by dry milling, which aims to achieve maximum capital return per liter of bioethanol, or wet milling, which uses higher capital investments to yield valuable products from corn grain prior to bioethanol fermentation step. Both processes use amylase enzyme to breakdown the complex starch network to glucose, which in turn is employed to yield bioethanol. US is the largest producer of bioethanol in the world, 95% of which is produced from corn starch (Bothast and Schlicher 2005).

Wheat (*Triticum aestivum*) is a grass grown for its grain, and is the most produced food crop in the world in terms of area harvested. The wheat grain employed for bioethanol processing is dried to about 14% moisture content, milled into the starchy flour, which is enzymatically hydrolyzed to glucose, which in turn is fermented to bioethanol, similar to corn milling process (Mortimer et al. 2004).

2.2.3 Lignocellulosic Feedstocks

As first-generation bioethanol uses food crops, the feedstock functionality is restricted. Agriculture and forest residues, fast-growing trees and energy crops, which together represent world's most abundant and renewable resource, constitute the lignocellulosic biomass available for the production of bioethanol (Table 2.1) (Balat et al. 2008; Guo et al. 2015).

Currently, the most frequent lignocellulosic feedstocks employed for commercial bioethanol production are by-products or wastes of sugar-mediated or starch-mediated bioethanol processes and serve to advance the overall bioethanol yield of the plant (Mortimer et al. 2004). In some plants, they also are employed to partially replace fossil fuels used to supply energy for these processes, thereby reducing emissions and facilitating a higher RER. Such feedstocks include bagasses from sweet sorghum and sugarcane, straws from sugar beet and wheat, corn stover and sugar beet pulp (Thiruvengadathan 2017).

One beneficial approach under research to produce lignocellulosic biomass has been to use low quality marginal land unsuitable for cultivation of food crops, for growing cellulose-rich grasses, crops, and trees. Perennial energy crops and grasses, for example, miscanthus, napier grass switchgrass, energy cane (a fiber rich variety of sugarcane), giant reed, and shrub willow can achieve high bioethanol yields at low-costs. Biomass from fast-growing trees, for example, pine, black locust, eucalyptus, and hybrid popular can generate wood chips rich in hemicellulose and cellulose (Thiruvengadathan 2017; Balat et al. 2008; Bomgardner 2013).

Lignocellulosic biomass desire pretreatment prior to fermentation, which help reduce crystallinity and material size, providing easier admittance for hydrolysis. Different pretreatment approaches are being widely studied (Alvira et al. 2010; Refaat 2012).

2.3 Biochemical Reactions Involved in Transformation of Vegetative Biomass to Bioethanol

The biochemical reactions which may be involved in the transformation of vegetative biomass to bioethanol are illustrated below (Cardona and Sánchez 2007; Guo et al. 2015). Hydrolysis of cellulose, starch or similar polysaccharides to hexose sugars (fructose and glucose).

$$(C_6H_{10}O_5)_n + nH_2O \rightarrow nC_6H_{12}O_6$$

Hemicellulose hydrolysis to pentose sugars (arabinose, mannose, xylose, etc.)

$$(C_5H_8O_4)_n + nH_2O \rightarrow nC_5H_{10}O_5$$

Hydrolysis of sucrose to fructose and glucose in *Saccharomyces cerevisiae*, catalyzed by the enzyme invertase.

$$C_{12}H_{22}O_{11} + H_2O \rightarrow C_6H_{12}O_6 + C_6H_{12}O_6$$

Transformation of pentoses and hexoses to bioethanol via the following exothermic reactions (enthalpy of formation of bioethanol, $\Delta_f H^\circ = -278$ kJ/mol)

$$C_6H_{12}O_6 \rightarrow 2C_2H_5OH + 2CO_2C_5H_{10}O_5 \rightarrow 5C_2H_5OH + 5CO_2$$

The reaction of bioethanol production from hexose sugar has a maximum theoretical yield of 0.511 g bioethanol per g of glucose utilized. Industrial first-generation bioethanol processes operate at >90% of theoretical yield (Thiruvengadathan 2017; Gombert and Antonius 2015).

2.4 Overview of Industrial Scale Formation of Bioethanol

The following section describes the different production processes of bioethanol using starch, sugar, and lignocellulosic feedstock.

2.4.1 Formation of Bioethanol *via* Sugar-Mediated Feedstock

The bioethanol formation process begins with washing (with water), grounding, and milling the sugarcane. The cane juice (black treacle or molasses) is then employed to make bioethanol, while the baggase (solid-type waste of the juice extraction processes) can be employed to create electrical energy via steam. Unlike other feedstocks, transformation of simple sugars (viz. sucrose) into bioethanol does not desire enzymatic hydrolysis of the feedstock. After the decontamination process and regulating the PH, the black treacle is then introduced to a fermentation assembly where yeasts are employed to ferment the black treacle into bioethanol (Gebreyohannes 2010). The yeast (*Saccharomyces cerevisiae aka* Brewer's or Baker's yeast) is continuously isolated through centrifugation and recycled to the fermenter. Lastly, the sugarcane bioethanol from the fermentation assembly is distilled to raise its level of purity to around 96% weight of bioethanol (Cardona Alzate and Sanchez Toro 2006).

2.4.2 Formation of Bioethanol *via* Starch-Mediated Feedstock

The formation of bioethanol from corn desires breakdown or scarification of polysaccharides into fermentable sugar. To begin with, the grains of corn are washed with water and grounded into small fragment to expose the corn starch, which is then converted by milling process into a fine powder, which is employed in the fermentation process. The fine powder is mixed with H_2O to dissolve the enzymes (viz. *α-amylase*) that will breakdown it partly into minor fragments (Cardona Alzate and Sanchez Toro 2006). The mesh is heated at 120 to 150 °C to liquefy the starch. The temperature is then raised to 225 °C to breakdown the starch extra. The mesh is then cooled and the second enzyme (viz. *glucoamylase*) (which breakdowns starch into glucose) is added (Ahmed et al. 2006). Next, the glucose is introduced to the fermentation assembly. The process of fermentation takes around 50 h and it breakdowns glucose to bioethanol and CO_2. Bioethanol is then purified to eliminate the residual H_2O and is denaturized through introducing 2 to 5% gasoline in order to make bioethanol unhealthy for human drinking (Gebreyohannes 2010).

2.4.3 Formation of Bioethanol *via* Lignocellulosic-Mediated Biomass by Enzymatic Hydrolysis and Fermentation

Lignocellulosic biomass is the inexpensive and most abundant raw material for the formation of bioethanol. The foundations of lignocellulosic biomass include paper wastes, agricultural residues, and woods (Guffey and Wingerson 2002). They have a highly complicated structure which comprises of lignin (nearly 25% of dry weight), hemicelluloses (nearly 30% of dry weight), and cellulose (nearly 45% of dry weight) (Kenney et al. 2013).

Lignocellulosic biomass can be transformed into bioethanol in different techniques. Enzymatic hydrolysis followed by fermentation is the most popular approach, which can be employed to form bioethanol from lignocellulosic biomass. In this approach, the key challenge is the pretreatment step, where an enzyme is introduced to breakdown the cellulose to glucose (Cardona Alzate and Sanchez Toro 2006). In this step, reagents like aqueous HNO_3, H_2SO_4 or HCl can also be used to breakdown the cellulose. But, the difficulty of this process results in higher manufacturing costs as compared to processes that use starch or sugar as raw materials (Gebreyohannes 2010).

2.4.4 Formation of Bioethanol *via* Lignocellulosic-Mediated Biomass by Gasification and Fermentation

Bioethanol can be obtained by gasification of lignocellulosic biomass, followed by fermentation of the syngas (*aka* synthesis gas, producer gas or synthetic gas). In this approach, a gasifier is employed to transform the lignocellulosic material into a producer gas at high-temperature. The producer gas is then cooled and introduced to a bioreactor where the process of fermentation takes place. The transformation of producer gas to bioethanol takes place in the bioreactor through special strains of bacteria under anaerobic environment (environment with little or no available O_2). Lastly, the bioethanol produced from the bioreactor is isolated from H_2O using a molecular sieve column and a distillation column (Gebreyohannes 2010). The summary of the bioethanol formation process through producer gas fermentation is illustrated in Fig. 2.2.

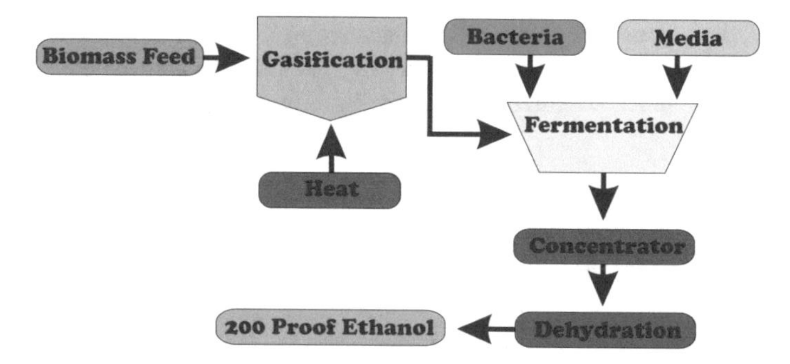

Fig. 2.2 Formation of bioethanol from lignocellulosic biomass

2.5 Recent Approaches for Sustainable Production of Bioethanol

2.5.1 Obtaining Bioethanol from Acid Pretreated Indian Bamboo Variety by Separate Hydrolysis and Fermentation

Bamboo is a fast-growing woody grass. In India, the total growing stock of bamboo is 80.4 MMT (Million Metric Tons) and the estimated annual harvest is 18.0 MMT. Usually for every 1 MT of bamboo produced around 0.3 MT residue in the form of top portion, branches, twigs, leaves, and roots (Singh and Gu 2010). In literature, bamboo (BM) was employed as a sustainable feedstock to obtain bioethanol after dilute acid pretreatment and enzymatic saccharification (Leenakul and Tippayawong 2010; Sathitsuksanoh et al. 2010). In 2013, R. Sindhu et al. produced bioethanol in yield of 1.758% (v/v) with an efficiency of 41.69%. In this investigation, different mineral and organic acids were screened to select the best pretreatment agent. Dilute H_2SO_4 was used by Raveendran Sindhu et al. and the efficiency of pretreatment was evaluated by enzymatic saccharification (Sindhu et al. 2014). Taguchi method was used to optimize the physical conditions such as the amount of biomass, concentration of acid, and incubation time. And it was experimental that the most operative physical parameters for pretreatment were almost 5% (w/w) acid concentration, 15% (w/w) biomass loading, and 30 min pretreatment time. And utilization of these favorable pretreatment conditions resulted in the production of 0.319 g/g of reducing sugar. Various parameters affecting enzymatic saccharification were deliberated by means of an experimental Box–Behnken design for response surface method. These parameters are enzyme loading, solid input, surfactant concentration, and time of incubation. The optimum conditions for hydrolysis were found to be 50 FPU of commercial cellulase enzyme, 11.25% (w/w) of biomass loading, as a surfactant Tween-80 in 0.125% (w/w) concentration, and 42 h of incubation. Under optimized enzymatic hydrolysis, 0.651 g/g of reducing sugar was synthesized. Fermentation of the enzymatic hydrolysed liquid from the pretreated biomass using *S. cerevisiae* showed bioethanol yield of 1.76% (v/v) with an efficiency of 41.69%. The results revealed that BM is a potential candidate for sustainable bioethanol production (Sindhu et al. 2014).

2.5.2 Production of Bioethanol from Sweet Sorghum Bagasse by Microwave Irradiation

Sweet sorghum (*Sorghum Bicolor L. Moench*) is grass specie that can grow even on a poor soil such as marginal land (Mastrorilli et al. 1999; Vasilakoglou et al. 2011). The growing time of this hardy crop is about six months. It is cultivated for its grain which has a high concentration of starch and is also used for human's food. Stalks of sorghum are rich in Lignocellulose materials and sugar syrup. Traces of sweet sorghum after juice and starch extraction are known as sorghum bagasse which has high Lignocellulose content. Bioethanol can be produced from sweet sorghum bagasse after passing it through various kinds of technologies such as pretreatment, hydrolysis, and fermentation, respectively (Vasilakoglou et al. 2011). The major difficulty to obtain bioethanol at large-scale from lignocellulose material is linked with elevated production cost which itself is associated with the use of expensive enzymes for hydrolysis and acids and bases for pretreatment. Sanette Marx et al. in 2013, reported a feasible and microwave-assisted production of sugar from grassy biomass, devoid of the utilization of enzymes, so that, sugar is produced from sweet sorghum bagasse after pretreatment and hydrolysis via single step employing microwave irradiation (Marx et al. 2014). A total sugar yield of 820 g kg^{-1} was obtained in a 50 g/kg H_2SO_4 solution in H_2O, with a power input of 43.2 kJ/g of dry biomass (i.e. 20 min at 180 W power setting). Bioethanol yield of 480 g per kilogram of total sugar was produced by fermentation in 24 h by means of mixed culture technique. A total sugar yield of 820 g kg^{-1} was attained in a 50 g kg^{-1} sulfuric acid solution in H_2O, with a power input of 43.2 kJ g^{-1} of dry biomass (i.e., 20 min at 180 W power setting). Bioethanol yield mediated on total sugar of 480 g kg^{-1} was attained after 24 h of fermentation using a mixed culture of organisms. These results illustrated that this process is efficient for economical production of bioethanol by grassy biomass, at a commercial scale such as 0.252 m^3 ton^{-1} or 33 m^3 ha^{-1} bioethanol, by using only lignocellulose part of the stalks. The cost problem was diminished by short pretreatment time, in addition to the fact that hydrolysis is done without expensive enzymes (Marx et al. 2014).

2.5.3 *Ultrasonic-Based Simultaneous SSF of Pretreated Oil Palm Fronds to Obtain Bioethanol*

Oil palm fronds is an agro-waste produced as residues after oil palm (*Elaeisguineensis Jacq.)* cultivation. About 70% of by-products from the palm oil industry constitute oil palm fronds biomass (OPF) (Effland 2000). To obtain sustainable bioethanol, these oil palm fronds referred to as second-generation feedstocks could replace the edible energy crops, which are currently employed to obtain bioethanol. In 2013, Cynthia Ofori-Boateng and Keat-Teong Lee, investigated ultrasonic-assisted bioethanol production from organosolv pretreated oil palm fronds (OPFs) via simultaneous saccharification and fermentation (SSF) (Ofori-Boateng and Lee 2014). Ultrasound application in industrial processes is able to reduce the process temperature and time, which eventually improves the energy utilization for sustainable production. Ultrasonic-assisted saccharification and fermentation (SSF) of pretreated OPFs was considered as the most promising method for significant bioethanol production in a very short time, as compared to that of SSF processes devoid of ultrasound application. Following are optimal conditions for maximum bioethanol production (18.2 g/l with yield 57.0%); 15 g/l of yeast, solid loading of 10% (w/v), 40 °C temperature, and reaction time of 5 h. The results demonstrated that the ultrasonic-assisted SSF is a sustainable approach to obtain bioethanol with high concentration from organosolv/H_2O_2 pretreated OPFs (Ofori-Boateng and Lee 2014).

2.5.4 *Low Energy Consumption Integrated Method to Transform Sucrose and Homocelluloses in Sweet Sorghum Stalks into Bioethanol*

For the purpose of production of bioethanol, sweet sorghum is considered as one of the propitious energy crops because it offers sugar, grain, as well as lignocellulosic material. In order to achieve reliable and cost-effective bioethanol production, it is essential that all of the carbohydrates in stalks (such as sucrose and simple fermentable sugars produced from lignocelluloses hydrolysis) must undergo biotransformation. In 2013, Jihong Li et al. described a new method for bioethanol production from sweet sorghum, based on low energy utilization (Li et al. 2013). This process was characterized by distillation along with pretreatment by alkali. Almost 10% (w/w DM) loading of NaOH was reported as optimum alkali concentration for pretreatment along with distillation step. In this research 91.9 kg of bioethanol/tone of fresh sweet sorghum stalk was produced under optimal conditions: enzyme loading of 10 FPU/g of glucan and 72 h hydrolysis time. This novel process is efficient to reduce the cost of bioethanol production and implement biotransformation of all carbohydrates in sweet sorghum stalks. In this novel investigation, pretreatment along with first-generation bioethanol distillation step is carried out in one reactor via single step, so that the pretreatment reactor cost was also saved. By means of this technology, production of fermentable sugar from lignocellulose biodegradation was made feasible by annihilation of obstinacy of lignocelluloses. Low energy-consumption integrated technology is a vital technology for the production of various products, for instance, butanol, sugar-based lignocellulosic resource, and biogas from bioconversion of sweet sorghum bagasse (Li et al. 2013).

2.5.5 *Usefulness of Low-Intensity Pulsed Ultrasound to Obtain Bioethanol*

M. Shaheen et al. studied that microbial metabolic activity can be enhanced by low-intensity pulsed ultrasound (LIPUS) technology which customizes bioethanol production by lignocellulosic. M. Shaheen et al. illustrated specific favorable conditions for LIPUS to improve metabolic potency of microorganisms such as *Trichoderma reesei* Rut C-30: fungus for cellulose degradation and *Saccharomyces cerevisiae:* yeast producing bioethanol (Shaheen et al. 2013). Conditions for LIPUS are as follows: ultrasounds of 1.5 MHz with the intensity of 80 mW/cm^2, 20% duty cycle, 5 min exposure with 12 exposures/day. LIPUS technology increases the production of cellulase by *T. reesei* up to 16 $\pm$ 6%. This methodology also causes an enhancement of 31 $\pm$ 10% in bioethanol production through *S. cerevisiae* which results in a collective improvement of 52 $\pm$ 16% in lignocellulosic production of bioethanol via LIPUS. So, LIPUS technology is a promising approach for improved production of bioethanol by lignocelluloses (Shaheen et al. 2013).

2.5.6 *Production of Bioethanol from Water Hyacinth Eichhornia Crassipes*

Eichhornia crassipes is the scientific name of water hyacinth, which is a hydrophyte (aquatic plant). It is native to South America but has been planted in temperate and tropical regions. It helps in the remediation of contaminated water (Hasan et al. 2007; Fang et al. 2007). *Eichhornia crassipes* has a very high growth rate which may assists its capability as raw material to obtain bioethanol. By means of various techniques, bioethanol can be produced by water hyacinth. In 2012, to determine optimum bioethanol production conditions, water hyacinth was saccharified with 3% (v/v) H_2SO_4 and subjected to cellulase treatments by Toshiyuki Takagi and co-workers (Takagi et al. 2012). The original saccharified solution yielded both glucose at 5.3 $\pm$ 0.2 g/l and reducing sugars at 9.7 $\pm$ 0.1 g/l. The concentration of the saccharified solution under vacuum at 70 °C yielded both glucose at 21.5 $\pm$ 2.9 g/l and reducing sugars at 33.3 $\pm$ 2.1 g/l (Arshad et al. 2008, 2011, 2014, 2017; Arshad 2010). An aquatic yeast *Saccharomyces cerevisiae* TY2 produced bioethanol at 9.6 $\pm$ 1.1 g/l from the concentrated saccharified solution. These results suggest that refinement of the saccharification process and application of the selected yeast could advance the efficiency of obtaining bioethanol from the water hyacinth (Takagi et al. 2012).

2.5.7 Bioethanol Production *via* Fermentation of Lemon Peel Wastes Pretreated with Steam Explosion

Talebnia et al., in 2008, demonstrated that citrus peel wastes (CPW) such as Lemon citrus peel wastes, can be employed to feed farm animals after dehydration or ensilage (Talebnia et al. 2008). It is investigated that lemon CPW is an agricultural waste rich in pectin and carbohydrates (Boluda-Aguilar et al. 2010; Marín et al. 2007; Mielenz et al. 2009). Further research by M. Boluda-Aguilar and López-Gómez revealed that bioethanol can be produced by lemon CPW by the means of steam explosion followed by enzymatic hydrolysis pretreatments on lemon (*Citrus limon L.*) (Boluda-Aguilar and López-Gómez 2013). These CPWs can also be utilized to produce bioethanol, as well as galacturonic acid, citrus pulp pellets, and d-limonene by simultaneous treatment of steam-exploded lemon peel wastes with hydrolysis and fermentation. The efficiency of this process was evaluated by measuring the amount and concentration of all primary and secondary products such as bioethanol, sugars, and galacturonic acid. Almost 60 L bioethanol/1000 kg fresh lemon CPW was produced when steam-exploded lemon CPW undergoes SSF processing by the use of minimum enzymatic concentration. This process is applicable at the laboratory scale, as well as commercial scale, in 5 L bioreactor. All these results disclosed that lemon CPW may be considered as a potential source to obtain bioethanol (Boluda-Aguilar and López-Gómez 2013).

2.5.8 Sono-Mediated Enzymatic Saccharification of Sugarcane Bagasse to Obtain Bioethanol

In 2012, a new synthetic approach for bioethanol production was introduced by Rajendran Velmurugan and Muthukumar, which is concerned with the utilization of sono-mediated pretreatment and enzymatic saccharification for production of bioethanol from sugarcane bagasse (SCB) (Velmurugan and Muthukumar 2012). Research work disclosed that sono-mediated alkali (sodium hydroxide) pretreatment effects removal of some polysaccharides (such as hemicellulose) and lignin from sugarcane bagasse and consequences of this investigation showed that about 90.6% of lignin and 80.8% of hemicelluloses was removed. Sono-mediated enzymatic saccharification was carried out with *Cellulomonas flavigena* (MTCC 7450) and the yield was observed to be influenced by pH, cell mass and liquid-to-solid ratio (LSR). The optimum conditions were found to be cell mass of 15 g/L, LSR of

15:1, pH of 6.0, and reaction time of 360 min. Under these optimum conditions, the theoretical yield of about 91.28% (38.4 g/L) of glucose was obtained. The use of ultrasounds may enhance the performance of the process due to cellulose swelling and faster enzymatic saccharification. *Zymomonas mobilis* (MTCC 89) was used for fermentation of the obtained hydrolyzate and around 91% of the theoretical yield of bioethanol was observed in 36 h of fermentation (Velmurugan and Muthukumar 2012).

2.5.9 *Sugarcane Bagasse Hydrolysis Approach Using Yeast Cellulolytic Enzymes*

For rational advancement, bioethanol production using lignocellulosic biomass is considered as the most promising technology. Cohesive combination among the biomass components develop resistance for hemicelluloses and cellulose hydrolysis to produce fermentable sugars, so it is significant to diminish chemical and physical barriers related to these cohesive forces. This work illustrated that yeast of Amazon region and Brazilian Cerrado biome has potential to hydrolyze sugarcane bagasse pretreated with sulfuric acid (De Souza et al. 2013). From Amazon regions and Minas Gerais Cerrado, almost 214 and 103 yeasts isolates were collected, among these yeast isolates 18 (17.47%) from Minas Gerais Cerrado and 11 (5.14%) from Amazon regions were taken as the cellulase source. A considerable amount of β-glucosidase was produced by ubiquitous *Cryptococcus laurentii* as compared to that produced from exo- and endo-glucanase activities. In *natura* sugarcane bagasse was pre-treated with 2% sulfuric acid for 0.5 h at 150 °C. This pretreatment produces fibrous residue which undergoes hydrolysis by means of enzyme extract of *Cryptococcus laurentii* yeast, for a time duration of 72 h. By enzymatic hydrolysis reaction of enzyme extract, about 32% of the cellulose (including 2.4% of glucose) was subjected to conversion, showing that *C. laurentii* is a potential source of β-glucosidase. The results imply the significance of isolating microbial strains that yield enzymes of biological interest, given their broad usefulness in the production of biofuel (De Souza et al. 2013).

2.5.9.1 Pretreatment of Solids from Corn Stover and Reed Straw Pretreated with Hot Water for Bioethanol Production

Pretreatment of lignocellulosic materials by liquid hot water (LHW) is an efficient and sustainable approach for bioethanol production. Efficient yield with high bioethanol content was produced from water insoluble solids (WIS) of corn stover and reed straw by LHW pretreatment utilizing a propitious technology known as fed-batch semi-SSF (Lu et al. 2013). But, this process requires a high amount of cellulose and wash water which restrains its practical application. To cope with these restrictions, WIS was pretreated with Tween 40 before bioethanol fermentation, which is considered as the most efficient and economical method for bioethanol fuel production from unwashed WIS produced from corn stover and reed straw, after LHW-assisted-pretreatment of WIS. Tween 40 pretreatment might trim down cellulose loading and amount of water wash. Liquid by-products of Tween 40 pretreatment can be reprocessed. Maximum amount of 52.26 g L^{-1} from corn stover and 56.28 g L^{-1} from reed straw was obtained by unwashed WIS after Tween 40 pretreatment and fermentation with fed-batch S-SSF under following optimum conditions: 1.5% concentration of Tween 40, WIS-to-Tween 40 ratio of 1:10 (w/v) and pretreatment time of one hour at the appropriate temperature. Calculated yields were 71.1 and 69.1% for corn stover and for reed straw, respectively (Lu et al. 2013).

2.5.9.2 Combine Acid and Gamma Rays Application to Enhance Enzymatic Hydrolysis of Wheat Straw

Lignocellulosic biomass is considered as renewable and economical raw material for fermentable sugars by bioconversion, which is further used for biofuel production. It is difficult to hydrolyze lignocelluloses without pretreatment due to presence crystalline cellulose, lignin, and hemicelluloses (Niu et al. 2009). Hong et al., reported a combined dilute acid and gamma irradiation pretreatment to improve enzymatic hydrolysis for the synthesis of bioethanol (Hong et al. 2014). This combined pretreatment using 3% H_2SO_4-gamma irradiation, significantly affects glucose production as compared to individually pretreated and untreated wheat straw. The overall increase in enzymatic hydrolysis is associated with the degradation of the crystalline framework of cellulose, removal of hemicelluloses, and lignin or lignin modification. Investigation showed that enzymatic hydrolysis of wheat straw biomass is enhanced by combined pretreatment (Hong et al. 2014).

2.5.9.3 Effect of Dual Salt on the Pretreatment of Sugarcane Bagasse with Peroxide

Sugarcane bagasse rich in lignocellulosic material is an agro-waste, some industries such as sugar mills also produce sugarcane bagasse as a by-product. In 2014, the effect of various dual salts on H_2O_2-assisted pretreatment of sugarcane bagasse (SCB) for bioethanol production was studied by Ramadoss and Muthukumar (2015). The results displays better cellulose recovery and delignification for zinc oxide and manganese sulfate monohydrate system. The delignification of 74.18 $\pm$ 1.51% and maximum cellulose recovery of 93.42 $\pm$ 1.74% were obtained at optimized conditions (1 g SCB, 1% H_2O_2, 30 min, 100 °C and 0.5:100 g/mL). SCB was passed through hydrolysis by using dilute acid, under most effective reaction conditions, arabinose, xylose, and glucose were produced in the concentration of 3.58 $\pm$ 0.17 g/L, 11.62 $\pm$ 0.21 g/L, and 15.26 $\pm$ 0.37 g/L, respectively. Analysis of XRD, SEM, and FTIR results highlights the importance of the given method for effective lignin removal and hydrolysis via destroying cellulosic crystal structure. The fermentation of hydrolyzate with

Saccharomyces cerevisiae displayed maximum bioethanol production of 13.1 g L^{-1} after 3 days and bioethanol yield of 0.184 g/g sugarcane bagasse (Ramadoss and Muthukumar 2015).

2.5.9.4 Sodium Hydroxide Pretreated Sugarcane Bagasse Using Phlebia Sp. MG-60 to Obtain Bioethanol

In 2014, Kondo et al. optimized sugarcane bagasse alkaline pretreatment for consolidated bioprocessing fermentation through *Phlebia sp.* MG-60; a fungus for cellulose fermentation (Kondo et al. 2014a, b). The xylan and lignin contents of sugarcane bagasse were reduced while a high bioethanol production, from *Phlebia sp.* MG-60 assisted pretreated bagasse, was made possible, in an alkaline concentration-dependent manner. The fungus afforded xylanase and cellulase rapidly over 5 days. 4.5 g/L bioethanol was produced when this fungus was cultured with 20 g/L of sugarcane bagasse pretreated with sodium hydroxide (60 min, 121 °C, 0.8 wt%). This bioethanol production is equivalent to 210 mg bioethanol per gram of the original untreated bagasse after 10 days fermentation, delivering bioethanol yields of 65.7% of the theoretical maximum. This investigation highlighted the significance of *Phlebia sp.* MG-60 for reliable bioethanol production via NaOH-pretreated bagasse in single bioreactor devoid of chemical and enzymatic hydrolysis (Kondo et al. 2014a, b).

2.5.9.5 Integrated Fungal Fermentation of Sugarcane Bagasse for Production of Bioethanol by Phlebia Sp. MG-60

Integrated fungal fermentation (IFF) involves bioethanol production by biological delignification by consolidated biological processing by *Phlebia* sp. (a white-rot fungus). Sugarcane bagasse was treated with MG-60. Biological delignification (removal of lignin by an enzyme) of bagasse is affected by initial moisture content and it was observed that 75% moisture is favorable, selectively, for degradation of lignin and simultaneous bioethanol production (Kondo et al. 2014a, b). Some additives, for example, minerals, organic compounds, and basal bodies also influence this delignification process. Various inorganic chemical reagents, for example, Cu^{2+}, Mn^{2+} or Fe^{2+}, reduced bagasse carbohydrate degradation via MG-60 during delignifying incubations and leads to increased production of bioethanol. Investigation reveals that appropriate culture conditions might enhance the efficiency of the IFF process (Kondo et al. 2014a, b).

2.5.9.6 Production of Bioethanol by Mangrove-Derived Marine Yeast

In 2013, Saravanakumar et al. disclosed the production of bioethanol from mangrove-derived marine yeast strain of *S. cerevisiae*. The marine strain of *S. cerevisiae* displayed the maximum bioethanol formation of 69.58% under the optimum conditions of temperature (30 °C), sawdust concentration of (6.85 mgl^{-1}) under the agitation speed of (360 rpm) in four days of incubation (Saravanakumar et al. 2013).

2.5.9.7 Production of Bioethanol from Waste of Date-Palm Fruit Using Solar Energy

In 2016, Boulal et al. developed a highly efficient solar energy-based approach for production of bioethanol from date palm waste, which is rich in biodegradable sugars. In this technique, a bioconversion system was designed and constructed using a butane gas distiller using a cocotte (cooker) of 30 L capacity and a solar batch fermenter (SBF) of 50 L capacity (Boulal et al. 2016). The bioconversion systems led to the formation of a large quantity of bioethanol at 90° after distillation of the date palm waste juice at 78 °C.

2.6 Conclusion

Renewable energy is core to the implementation of SDG 7 (ensure access to reliable, affordable, sustainable, and modern energy for all). New developments in renewable energy are essential to replace fossil fuels in the global energy system, and can also bring modern, cheap energy to the almost 1.1 billion around the world who lack access to electrical energy. Access to clean energy is also a necessary requirement to attain many of the other SDGs. Among all renewable energy sources, the biomass energy source is one of the most effective sources because its supplies are not limited. We can always grow crops and trees, and people will always produce garbage. Apart from the unlimited supply, it is a highly affordable and reliable source of energy. Moreover, increasing the use of biomass leads to the reduction of greenhouse gas emissions, better maintenance of forests, and security of energy supply and increasing the productivity of forests. Biomass has the highest potential among all the other green energy sources. Unlike other renewable energy sources, biomass can be transformed directly into liquid fuels-biofuels for our transportation requirements (trains, airplanes, busses, trucks, and cars). The most common type of biofuel obtained from biomass is bioethanol. However, the bioethanol production from biomass is an extremely important and challenging task because all the affordability, reliability, and suitability depends on the efficiency of bioethanol production.

Over the past few decades, researches are developing highly effective technologies to advance the process of bioethanol production from different biomass feedstocks in order to implement SDG-7. This chapter deals with the formation of bioethanol from different biomass feedstocks-Indian bamboo, sweet sorghum bagasse, oil palm, sweet sorghum stalk, water hyacinth, lemon peel, sugarcane bagasse, corn stover, reed straw, and wheat straw-using modern technologies including sono-mediated enzymatic saccharification, microwave irradiation, ultrasonic irradiation, steam explosion, and gamma irradiation. These modern technologies for bioethanol production from different biomass feedstocks offer reliable systems to achieve sustainable and profitable energy and healthier environments. Summarily, implementation of modern technologies to produce bioethanol can have a large influence on the energy system. First, it can help lower the environmental impacts of the energy system, assuming bioenergy is produced in a sustainable way with limited impact on land-based carbon stocks. Secondly, energy prices can be affected by increased competition with other (renewable) energy sources.

The goal of this chapter is to increase the use of biomass for energy production in an economic competitive and sustainable manner with the least impact on the environment; to close the energy access gap; and to address energy challenges including low quantity, quality, and reliability of the power supply and implementation of and compliance with regulations, as well as affordability of energy. In this chapter, pros and cons of each sustainable approach for bioethanol production are discussed in detail, which will be very fruitful for the research community to get knowledge about the finest technologies, to improve traditional methods and to develop further new methodologies for bioethanol production. Also, this chapter is highly helpful to identify and analyze critical issues, barriers, and opportunities that will help in the implementation of SDG-7.

References

Ahmed A, Cateni BG, Huhnke RL, Lewis RS (2006) Effects of biomass-generated producer gas constituents on cell growth, product distribution and hydrogenase activity of Clostridium carboxidivorans P7T. Biomas Bioenerg 30:665–672

Akbulut M, Özcan MM (2008) Some physical, chemical, and rheological properties of sweet sorghum (Sorghum Bicolor (L) Moench) Pekmez (Molasses). Int J Food Prop 11:79–91

Akorede MF, Hizam H, Pouresmaeil E (2010) Distributed energy resources and benefits to the environment. Renew Sustain Energ Rev 14:724–734

Alvira P, Tomás-Pejó E, Ballesteros MJ, Negro MJ (2010) Pretreatment technologies for an efficient bioethanol production process based on enzymatic hydrolysis: a review. Biores Tech 101:4851–4861

Andrzejewski B, Eggleston G, Lingle S, Powell R (2013) Development of a sweet sorghum juice clarification method in the manufacture of industrial feedstocks for value-added fermentation products. Ind Crop Prod 44:77–87

Arshad M (2010) Bioethanol: a sustainable and environment friendly solution for Pakistan . A Sci J COMSATS–Sci Vision 2011:16–17

Arshad M, Khan ZM, Shah FA, Rajoka MI (2008) Optimization of process variables for minimization of byproduct formation during fermentation of blackstrap molasses to ethanol at industrial scale. Lett Appl Microbiol 47:410–414

Arshad M, Zia MA, Asghar M, Bhatti H (2011) Improving bio-ethanol yield: using virginiamycin and sodium flouride at a Pakistani distillery. Afr J Biotech 10:11071

Arshad M, Ahmed S, Zia MA, Rajoka MI (2014) Kinetics and thermodynamics of ethanol production by Saccharomyces cerevisiae MLD10 using molasses. App Biochem Biotech 172:2455–2464
Arshad M, Hussain T, Iqbal M, Abbas M (2017) Enhanced ethanol production at commercial scale from molasses using high gravity technology by mutant S. cerevisiae. Braz J Microbiol 48:403–409
Astolfi-Filho Z, Telis VRN, de Oliveira EB, dos Reis CJS, Telis-Romero J (2011) Rheology and fluid dynamics properties of sugarcane juice. Biochem Engine J 53:260–265
Balat M, Balat H (2009) Recent trends in global production and utilization of bio-ethanol fuel. App Energ 86:2273–2282
Balat M, Balat H, Öz C (2008) Progress in bioethanol processing. Prog Energ Comb Sci 34:551–573
Barcelos CA, Maeda RN, Santa Anna LMM, Pereira N (2016) Sweet sorghum as a whole-crop feedstock for ethanol production. Biomas Bioenerg 94:46–56
Boluda-Aguilar M, García-Vidal L, González-Castaneda FP, López-Gómez A (2010) Mandarin peel wastes pretreatment with steam explosion for bioethanol production. Bioreso Tech 101:3506–3513
Boluda-Aguilar M, López-Gómez A (2013) Production of bioethanol by fermentation of lemon (Citrus limon L.) peel wastes pretreated with steam explosion. Ind Crops Prod 41:188–197
Bomgardner MM (2013) Seeking Biomass Feedstocks That Can Compete. Chem Eng News 91:11–15
Bothast RJ, Schlicher MA (2005) Biotechnological processes for conversion of corn into ethanol. App Microbio Biotech 67:19–25
Boulal A, Kihal M, Khelifi C, Benali B (2016) Bioethanol production from date palm fruit waste fermentation using solar energy. Afr J Biotechnol 15:1621–1627
Bozell JJ (2008) Feedstocks for the future–biorefinery production of chemicals from renewable carbon. Clean (Weinh) 36:641–647
Brundtland GH (1987) Our Common Future, 3rd edn. United Nations World Commission on Environment and Development, Oxford University Press, New Delhi
Cao JF, Gao BP, Gu WB (2006) Study on Producing Alcohol Fermentation Conditions by Sweet Sorghum Juice. Acta Agric Bor Sin 15:201–203
Cardona Alzate CA, Sanchez Toro OJ (2006) Energy consumption analysis of integrated flowsheets for production of fuel ethanol from lignocellulosic biomass. Energy 31:2111–2123
Cardona CA, Sánchez ÓJ (2007) Fuel ethanol production: process design trends and integration opportunities. Biores Tech 98:2415–2457
Chum HL, Warner E, Seabra JE, Macedo IC (2014) A comparison of commercial ethanol production systems from Brazilian sugarcane and US corn. Biofuels Bioprod Bioref 8:205–223
Coelho DT, Dale RF (1980) An energy-crop growth variable and temperature function for predicting corn growth and development: planting to Silking 1. Agronomy J 72:503–510
De Souza AC, Carvalho FP, Batista CFS, Schwan RF, Dias DR (2013) Sugarcane bagasse hydrolysis using yeast cellulolytic enzymes. J Microbio Biotech 23:1403–1412
Deesuth O, Laopaiboon P, Jaisil P, Laopaiboon L (2012) Optimization of nitrogen and metal ions supplementation for very high gravity bioethanol fermentation from sweet sorghum juice using an orthogonal array design. Energies 5:3178–3197
Demirbas A (2009) Biofuels securing the planet's future energy needs. Energ Convers Man 50:2239–2249
Demirbas A, Karslioglu S (2007) Biodiesel production facilities from vegetable oils and animal fats. Energ Sou a 29:133–141
Dodić JM, Vučurović DG, Dodić SN, Grahovac JA, Popov SD, Nedeljković NM (2012) Kinetic modelling of batch ethanol production from sugar beet raw juice. Appl Energ 99:192–197
Effland AB (2000) US farm policy: the first 200 years. Agric Outlook 269:5–21
Ergun M, Mutlu SF (2000) Application of a statistical technique to the production of ethanol from sugar beet molasses by Saccharomyces cerevisiae. Biores Tech 73:251–255

Fang YY, Yang XE, Chang HQ, Pu PM, Ding XF, Pu PM, Rengel Z (2007) Phytoremediation of nitrogen-polluted water using water hyacinth. J Plant Nutr 30:1753–1765
Gebreyohannes S (2010) Process design and economic evaluation of an ethanol production process by biomass gasification. 1st ed. Oklahoma State University, ProQuest Dissertations Publishing
Ghosh P, Ghose TK (2003) Bioethanol in India: recent past and emerging future. Adv Biochem Eng/biotech 85:1–27
Gombert AK, Antonius JA (2015) Improving conversion yield of fermentable sugars into fuel ethanol in 1st generation yeast-based production processes. Curr Opi Biotech 33:81–86
Guffey FD, Wingerson RC (2002) Fractionation of lignocellulosic biomass for fuel-grade ethanol production. Nat Energ Tech Lab. https://doi.org/10.2172/807155
Guo M, Song W, Buhain J (2015) Bioenergy and biofuels: History, status, and perspective. Renew Sust Energ Rev 42:712–725
Gurung A, Oh SE (2013) Conversion of traditional biomass into modern bioenergy systems: a review in context to improve the energy situation in Nepal. Renew Energ 50:206–213
Hasan SH, Talat M, Rai S (2007) Sorption of cadmium and zinc from aqueous solutions by water hyacinth (Eicchornia crassipes). Bioresour Technol 98:918–928
Hong SH, Lee JT, Lee S, Wi SG, Cho EJ, Singh S, Lee SS, Chung BY (2014) Improved enzymatic hydrolysis of wheat straw by combined use of gamma ray and dilute acid for bioethanol production. Rad Phy Chem 94:231–235
John RP, Anisha GS, Nampoothiri KM, Pandey A (2011) Micro and macroalgal biomass: a renewable source for bioethanol. Bioresour Technol 102:186–193
Kenney KL, Smith WA, Gresham GL, Westover TL (2013) Understanding biomass feedstock variability. Biofuel 2013(4):111–127
Kim M, Donal FD (2011) Composition of sugar cane, energy cane, and sweet sorghum suitable for ethanol production at louisiana sugar mills. J Ind Microbiol Biotechnol 38:803–807
Kondo R, De Leon R, Anh TK, Meguro S, Shimizu K, Kamei I (2014a) Effect of chemical factors on integrated fungal fermentation of sugarcane bagasse for ethanol production by a white-rot fungus, Phlebia sp. MG-60. Bioreso Tech 167:33–40
Kondo R, De Leon R, Anh TK, Shimizu K, Kamei I (2014b) Bioethanol production from alkaline-pretreated sugarcane bagasse by consolidated bioprocessing using Phlebia sp. MG-60. Int Biodeterior Biodegradation 88:62–68
Kumar CG, Rao RN, Srinivasa Rao P, Kamal A, Ashok Kumar A, Ravinder Reddy C, Reddy BVS (2013) Assessing sweet sorghum juice and syrup quality and fermentation efficiency. Int Crop Res Inst Semi-Arid Trop 2013:172–192
Laluce C, Leite GR, Zavitoski BZ, Zamai TT, Ventura R (2016) Fermentation of sugarcane juice and molasses for ethanol production. In: Ian O, Sagadevan M (eds) Sugarcane-based biofuels and bioproducts. Willey Hoboken, New York p, pp 55–86
Laopaiboon L, Nuanpeng S, Srinophakun P, Klanrit P, Laopaiboon P (2009) Ethanol production from sweet sorghum juice using very high gravity technology: effects of carbon and nitrogen supplementations. Biores Tech 100:4176–4182
Leenakul W, Tippayawong N (2010) Dilute acid pretreatment of bamboo for fermentable sugar production. J Sustain Energ Environ 1:117–120
Li J, Li S, Han B, Yu M, Li G, Jiang Y (2013) A novel cost-effective technology to convert sucrose and homocelluloses in sweet sorghum stalks into ethanol. Biotech Biofuel 6:174
Lu J, Li X, Yang R, Zhao J, Qu Y (2013) Tween 40 pretreatment of unwashed water-insoluble solids of reed straw and corn stover pretreated with liquid hot water to obtain high concentrations of bioethanol. Biotech Biofuel 6:159
Marín FR, Soler-Rivas C, Benavente-García O, Castillo J, Pérez-Alvarez JA (2007) By-products from different citrus processes as a source of customized functional fibres. Food Chem 100:736–741
Marx S, Ndaba B, Chiyanzu I, Schabort C (2014) Fuel ethanol production from sweet sorghum bagasse using microwave irradiation. Biomass Bioenergy 65:145–150

Mastrorilli M, Katerji N, Rana G (1999) Productivity and water use efficiency of sweet sorghum as affected by soil water deficit occurring at different vegetative growth stage. Eur J Agron 11:207–215
Matsuoka Y, Vigouroux Y, Goodman MM, Sanchez J, Buckler E, Doebley J (2002) A single domestication for maize shown by multilocus microsatellite genotyping. Proc Nat Acad Sci 99:6080–6084
Melorose J, Perroy R, Careas S (2015) World population prospects: the 2015 revision- key findings and advance tables. Pop Div 1:61–66
Menon V, Rao M (2012) Trends in bioconversion of lignocellulose: biofuels, platform chemicals & biorefinery concept. Prog Energ Comb Sci 38:522–550
Mielenz JR (2001) Ethanol production from biomass: technology and commercialization status. Curr Opin Microbiol 4:324–329
Mielenz JR, Bardsley JS, Wymana CE (2009) Fermentation of soybean hulls to ethanol while preserving protein value. Bioreso Tech 100:3532–3539
Moriarty P, Honnery D (2017) Assessing the climate mitigation potential of biomass. AIMS Energy 5:20–38
Mortimer ND, Elsayed MA, Horne RE (2004) Energy and greenhouse gas emissions for bioethanol production from wheat grain and sugar beet. Sheffield Hallam University, Sheffield, Resources Research Unit School of Environment and Development
Nigam PS, Singh A (2011) Production of liquid biofuels from renewable resources. Prog Energy Combust Sci 37:52–68
Niu K, Chen P, Zhang X, Tan WS (2009) Enhanced enzymatic hydrolysis of rice straw pretreated by alkali assisted with photocatalysis technology. J Chem Technol Biotechnol 84:1240–1245
Ofori-Boateng C, Lee KT (2014) Ultrasonic-assisted simultaneous saccharification and fermentation of pretreated oil palm fronds for sustainable bioethanol production. Fuel 119:285–291
Ogbonna JC, Mashima H, Tanaka H (2001) Scale up of fuel ethanol production from sugar beet juice using loofa sponge immobilized bioreactor. Biores Tech 76:1–8
Pacala S, Socolow R (2004) Stabilization wedges: solving the climate problem for the next 50 years with current technologies. Science 305:968–972
Pejin D, Mojović LJ, Vučurović V, Pejin J, Denčić S, Rakin M (2009) Fermentation of wheat and triticale hydrolysates: A comparative study. Fuel 88:1625–1628
Pérez-Lombard L, Ortiz J, Pout C (2008) A review on buildings energy consumption information. Energ Build 40:394–398
Ramadoss G, Muthukumar K (2015) Influence of dual salt on the pretreatment of sugarcane bagasse with hydrogen peroxide for bioethanol production. Chem Eng J 260:178–187
Refaat AA (2012) Biofuels from waste materials. Comp Renew Energ 5:217–261
Saravanakumar K, Senthilraja P, Kathiresan K (2013) Bioethanol production by mangrove-derived marine yeast, Sacchromyces cerevisiae. J King Saud Uni-Sci 25:121–127
Sathitsuksanoh N, Zhu Z, Hoc T, Bai M, Zhang YP (2010) Bamboo saccharification through cellulose solvent-based biomass pretreatment followed by enzymatic hydrolysis at ultra-low cellulase loadings. Bioresour Technol 101:4926–4929
Shafiee S, Topal E (2009) When will fossil fuel reserves be diminished? Energ Pol 37:181–189
Shaheen M, Choi M, Ang W, Zhao Y, Xing J, Yang R (2013) Application of low-intensity pulsed ultrasound to increase bio-ethanol production. Renew Energy 57:462–468
Sindhu R, Kuttiraja M, Binod P, Sukumaran RK, Pandey, (2014) Bioethanol production from dilute acid pretreated Indian bamboo variety (Dendrocalamus sp.) by separate hydrolysis and fermentation. Ind Crop Prod 52:169–176
Singh J, Gu S (2010) Biomass conversion to energy in India—a critique. Renew Sust Energ Rev 14:1367–1378
Smith P, Porter JR (2018) Bioenergy in the IPCC assessments. GCB Bioenergy 10:428–431
Staniszewski M, Kujawski W, Lewandowska M (2007) Ethanol production from whey in bioreactor with co-immobilized enzyme and yeast cells followed by pervaporative recovery of product–Kinetic model predictions. J Food Eng 82:618–625

Takagi T, Uchida M, Matsushima R, Ishida M, Urano N (2012) Efficient bioethanol production from water hyacinth Eichhornia crassipes by both preparation of the saccharified solution and selection of fermenting yeasts. Fish Sci 78:905–910

Talebnia F, Pourbafrani M, Lundin M, Taherzadeh MJ (2008) Optimization study of citrus wastes saccharification by dilute-acid hydrolysis. BioResources 3:108–122

Tanaka N (2011) Technology Roadmap-Biofuels for Transport. Int Energy Agency 1:1–56

Thiruvengadathan TN (2017) Bioethanol production using saccharomyces cerevisae cultivated in sugarcorn juice. In: Electronic Thesis and Dissertation Repository. The University of Western Ontario. https://ir.lib.uwo.ca/etd/4645

Vasilakoglou I, Dhima K, Karigiannides N, Gatsis T (2011) Sweet sorghum productivity for biofuels under increased soil salinity and reduced irrigation. Field Crops Res 120:38–46

Velmurugan R, Muthukumar K (2012) Sono-assisted enzymatic saccharification of sugarcane bagasse for bioethanol production. Biochem Eng J 63:1–9

Zabed H, Faruq G, Sahu JN, Azirun MS, Hashim R, Nasrulhaq Boyce A (2014) Bioethanol production from fermentable sugar juice. The Sci Wor J 2014:11

Chapter 3
Organic Waste Utilization for Sustainable Ethanol Production

Ali Akbar, Muhammad Bilal Sadiq, Imran Ali, and Muhammad Naeem

Abstract Every year tons of organic waste, coming from industries and agricultural, are going to land fill without proper utilization. These waste materials are mainly composed of energy-rich biomolecules such as lignocellulose, cellulose, pectin, carbohydrates, polysaccharides, sugars, proteins, fats, etc. Majority of these biomolecules are coming from food and non-food wastes, soft wood, hard wood, manures, agriculture crops such as sugarcane, bagasse, straw, vegetables, fruits, food, paper industries, etc. These organic waste materials can be utilized for sustainable development of bioethanol on large scale. The utilization of organic material for ethanol production can be achieved with and without pre-treatment procedures. This chapter is focused on the sustainable utilization of food and non-food organic waste materials for ethanol production. A thoughtful insight about sustainable ethanol production and processing is also part of this chapter.

Keywords Sustainability · Alcohol · Food waste · Non-food waste · Distilleries · Drinks

The demand for energy and fuel is increasing with the modernization. To meet today's need, it is essential to search for the alternate sources that can provide clean, affordable and sustainable energy. Organic waste and by-products from industries and agriculture can be a potential and sustainable source for providing the energy-rich molecules which can be further processed to obtain biofuel. The production of biofuel from organic waste not only reduces the environmental hazards but also it helps to produce sustainable and affordable energy that is in accordance with sustainable development goal 7 of the United Nations.

A. Akbar (✉) · M. Naeem
Department of Microbiology, University of Balochistan Quetta, Quetta, Pakistan
e-mail: aliakbar.uob@gmail.com

M. B. Sadiq
Department of Biological Sciences, Forman Christian College Lahore, Lahore, Pakistan

I. Ali
School of life Science and Engineering, Southwest University of Science and Technology, Mianyang, China

M. Arshad (ed.), *Sustainable Ethanol and Climate Change*,
https://doi.org/10.1007/978-3-030-59280-6_3

3.1 Introduction

The concept of green technology is gaining interest with an emphasis on utilizing the waste as raw material to produce valuable products (Sadiq et al. 2018). The two factors responsible for increasing demand of biofuel include population growth and extensive industrialization (Sarkar et al. 2012; Arshad et al. 2017). The global economy is greatly influenced by fossil energy sources, as these are the major sources of biofuel production (Uihlein and Schebek 2009; Arshad et al. 2018a). In urban areas, over consumption of fossil fuels resulted in high levels of pollution, and therefore to overcome this issue conventional fuel can be replaced with renewable biofuels such as bioethanol, biodiesel, and biohydrogen derived from sugarcane, corn, algae, etc. The importance of ethanol is gaining considerable interest due to global warming and climate changes associated with fossil fuel consumption (Arshad et al. 2014). Many countries shifted their focus toward the production of renewable source of biofuel such as bioethanol (Arshad et al. 2018b). United States and Brazil are the major ethanol-producing countries, accountable for 62% of ethanol of the world production (Kim and Dale 2004). Ethanol at industrial level is mainly produced in Brazil and USA from sugarcane and corn, respectively. However, bioethanol production from sugarcane and corn may not be sustainable solution due to feed value and serious concerns of food security (Nguyen et al. 2007).

The term bioethanol refers to ethanol derived from biological waste or renewable sources and can be used as fuel, chemical feedstock, and solvent in various industrial applications (Domínguez-Bocanegra et al. 2015). Bioethanol has several benefits over conventional fuel due to its acceptability, environmental friendly nature (for being less toxic and pollutive) and an alternative to fossil fuel (Arshad 2011). The bioethanol can be added to conventional fuel up to 5% without any modification (Ruiz et al. 2012). The renewable, nontoxic, and biodegradable nature of bioethanol accounts for it worldwide use as biofuel for transportation; moreover, it reduces the particulate emissions in engines due to oxygenated nature (Razmovski and Vučurović 2012).

The organic waste is one of the major environmental concerns and the utilization of agricultural and industrial waste as a source of biomass for generating energy can be an alternative and sustainable approach for developed and developing countries (Vučurovi et al. 2012). The food security and land use issues raised significant concerns over bioethanol production from feed crops (Paschalidou et al. 2016). The use of organic waste materials can be a promising source for biofuels as the first-generation fuels require feed crops (Nasterlack et al. 2014). However, production of biofuel requires advanced biotechnological applications to use organic waste as a source. The agro-industrial waste and animal manures present the diverse sources of substrates for bioethanol. The agro-industrial waste mainly includes protein-based waste, vegetable oils, animal manure, animal fats, organic feedstock, such as material composed of lignocellulose and animal slurries (Schievano et al. 2009). Bioconversion of solid waste is a green and sustainable method that produces high-value products such as biofuels and animal feedstocks by using microorganisms like *Pseudomonas*, *Bacillus*, etc. (Akinsemolu 2018). Biofuels are sustainable solution and promising alternatives to depleting natural resources of energy (Akinsemolu 2018).

Due to increasing fuel prices, energy supplies, and climate change concerns, lignocellulose biomass has gained considerable interest in biofuel production (González-García et al. 2013). The second-generation biofuels obtained from lignocellulosic biomass are environment friendly (González-García et al. 2012). Moreover, it is important for production of biofuel to have a source of biomass which is cost-effective, as the technological aspects are still at developmental stage (Arshad et al. 2011). The biomass accumulation and storage systems are essential for placement of fully equipped biofuel plant to process organic wastes. This approach can be implemented by improving the legislative policies for waste collection and storage (Stephen and Periyasamy 2018). The governments and policy-makers should subside for development of biofuels from organic wastes and provide incentive for utilizing biowaste-based fuels in order to maintain their presence consistent in energy market. United Nations' Sustainable Development Goals (SDGs) and Intergovernmental Panel on Climate Change (IPCC) provide guidelines for society, to cope with the leading challenges. Among these challenges, energy is predominant. Therefore, the SDGs have become dominant for energy policy-making (McCollum et al. 2018). In 2015, a comprehensive agenda has been developed by the member states of United Nations; Transforming our world: the 2030 *agenda for sustainable development*, more commonly known as the Sustainable Development Goals (UN 2015). The replacement of fossil fuels with renewable energy sources is one of the main goals of SDGs with the aim of climate change mitigation and reduction in carbon dioxide emissions (McCollum et al. 2018). In liaises with United Nationals Goal 7 (SDG 7) and Intergovernmental Panel on Climate Change (IPCC), several studies have been proposed as a model to produce sustainable, affordable, and clean energy. A farm in Tunisia has developed an alternative method to produce biofuel from firewood. This Chanouf farm designed a waste recycling unit to produce organic charcoal and biofuel from agroforestry waste (Rajab 2018).

The production of second-generation bioethanol is important and gaining interests at global level due to improved cost by using waste sources, doesn't interfere with food security issues and CO_2 balance (Dias et al. 2011). Therefore, production of bioethanol from organic waste can be an effective alternate for waste management and generation of second-generation fuel to meet the global fuel demand.

3.2 Bioethanol Utilization

Bioethanol is also referred to as ethyl alcohol (EtOH) or chemically C_2H_5OH. This could be utilized as pure ethanol directly or could also be combined with gasoline to get "gasohol" (Staniszewski et al. 2007). Bioethanol could be utilized as an octane enhancer or a gasoline improver and in diesel–bioethanol mixture to lessen the release of exhaust gases (Pejin et al. 2009). Globally, bioethanol is the most common biofuel

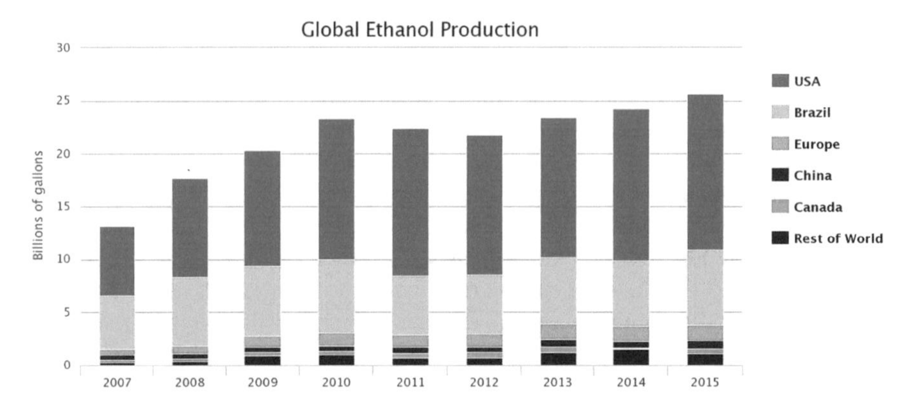

Fig. 3.1 Worldwide bioethanol production from year 2007 to 2015. (Adopted from: Renewable Fuels Association (2016) Available from: https://ethanolrfa.org/)

produced and prominently in the United States and Brazil two major bioethanol-producing countries accounting for 62% of the total world production. In Brazil, ethanol is manufactured on a large scale as fuel from sugar cane, whereas corn is used as a major source for bioethanol production in the United States (Hill et al. 2006). The worldwide bioethanol production from 2007 to 2015 by various countries is shown in Fig. 3.1. It could be observed that after a decline in 2011 and 2012, the bioethanol production reached highest in 2015. In addition, it depicts the two major countries such as USA and Brazil with maximum production.

3.3 Advantages of Bioethanol

Bioethanol is preferred due to higher flame speeds, broader flammability limits, higher octane number, and increased heat of vaporization (Balat and Balat 2009). Bioethanol is easily biodegradable and produces less environmental pollutants compared to fossil fuels (John et al. 2011).

3.4 Organic Waste Production and Utilization

The organic wastes can be categorized into waste materials from kitchen, garden, forest, farm animals, sewage sludge, municipal waste lignocellulose materials, etc. Production of bioethanol can be done by using the organic waste as raw material (Stephen and Periyasamy 2018). The organic waste materials can be categorized according to the location, as organic feedstocks of municipal waste are readily available in urban areas whereas agricultural wastes are available for production of biofuels in rural areas. The forests and lignocellulosic biomass are better options in areas of wood products processing due to their accessibility and cost benefits. Therefore, availability and cost are major driving forces to produce biofuels in various parts of the world (Arshad et al. 2008; Alberts et al. 2016; Stafford et al. 2017).

3.5 Organic Waste Utilization for Bioethanol Production

Approximately, 1391 million liters of bioethanol is produced in the world annually from organic waste, whereas the total global production of bioethanol is 118 billion liters. The major aims of biofuels production from organic wastes are improving production efficiency, sustainability, and emission control. To meet the need most of the innovative biofuel technologies are still under progression stage (IEA-ETSAP and IRENA 2013). The predominant technologies to convert organic wastes into biofuels include the fermentation of starch to ethanol and waste oil or animal fats transesterification to biodiesels. The organic waste requires various treatments to produce ethanol (Fig. 3.2).

3.5.1 Wood

Bioethanol production from wood materials by converting into sugar followed by microbial fermentation is gaining interest for being a sustainable source of energy (Zhang et al. 2016). The wood is abundant in natural polysaccharides that can be tailored into renewable energy by microbial fermentation (Saha et al. 2011). The cellulosic biomass is cheap and readily available, and therefore it can be the most suitable option to produce bioethanol without any interference with food security issues (Jung et al. 2015). The wood-based feedstock is rich in lignocellulose and considered as one of the most sustainable sources to produce second-generation bioethanol (Cai et al. 2017). The lignocellulose-based materials present complex structure in comparison to starch feedstocks; therefore, pre-treatment is essential to convert lignocellulose into sugar (Cunha et al. 2018). The pre-treatment alters the lignocellulose structure and increases the glucan and xylan availability for saccharification (Jung et al. 2015; Alvira et al. 2010). The hydrothermal pre-treatment is an efficient method for production of bioethanol from wood materials by simultaneous saccharification and fermentation (SSF) (Vargas et al., 2015; Domínguez et al. 2017). The pre-treatment converts the hemicellulose into cellulose and lignin, which determine the final ethanol yield.

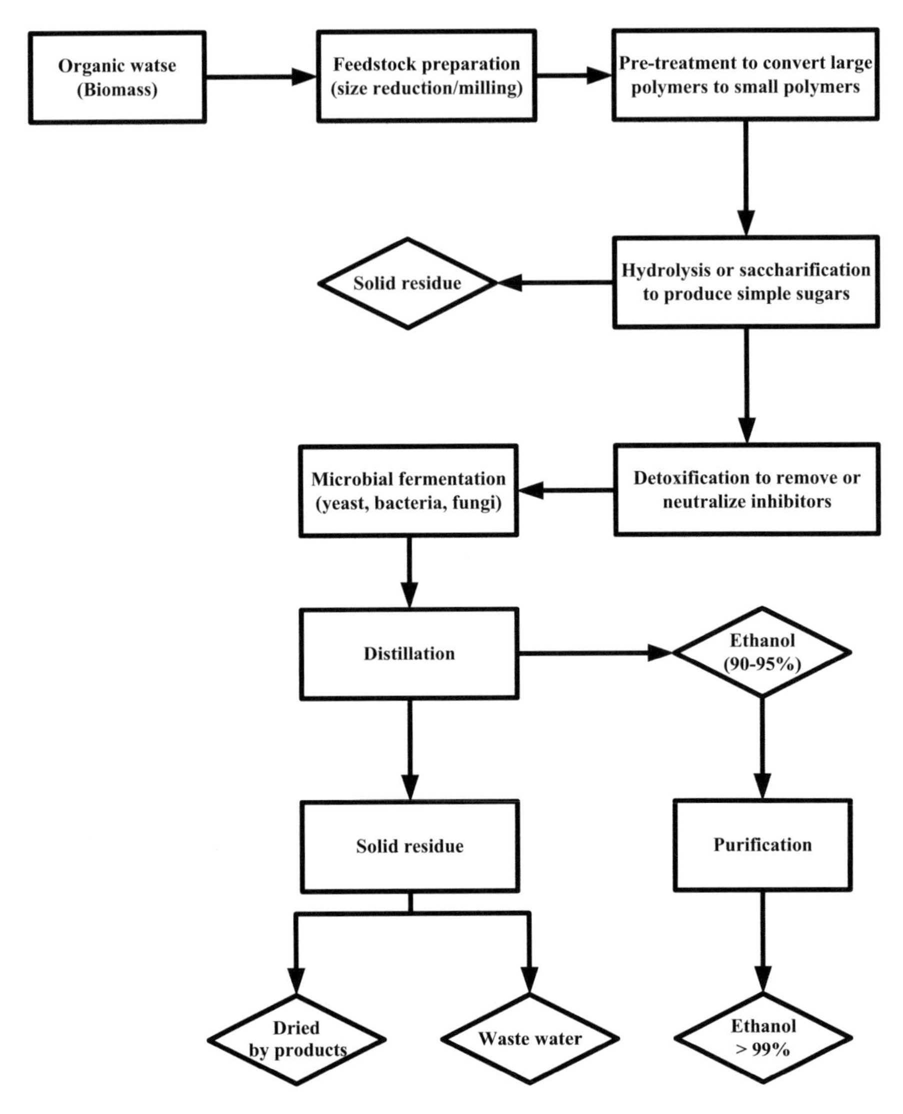

Fig. 3.2 Bioethanol production from organic waste (Adopted from Taherzadeh et al. 2013)

3.5.2 Manures

Tons of fecal wastes are produced annually by dairy cattle, goats, and broiler chicken, approximately 2372 kg/cattle and 48 kg/chicken of manure are produced each year (Chastain and Camberato 2004; Chastain et al. 2001). The production of waste in huge amount causes threats to environment by polluting air through greenhouse gas emissions (GHG) and water pollution (Miller et al. 2011; Davis et al. 2012). Whereas

10% of agrarian GHG releases in the environment are produced by manures alone (Steinfeld et al. 2006). As fertilizer the fecal waste has been used since ages, but in current scenario presence of antibiotics and pathogenic microorganisms restricts its applications as fertilizer (Hansen 2006; Venglovsky et al. 2009). Therefore, an alternative and efficient utilization of fecal waste could be solution to overcome the hazards associated with fecal waste. There is an urgent need for sustainable energy sources due to increasing world demand for energy. The fecal waste has documented evidence of being used to produce bioenergy (Kargbo 2010; Kim et al. 2014). Manures are rich in carbohydrates (30–50%) and 50–80% of which can be converted into simple sugars (Chen et al. 2005; Liao et al. 2006). This indicates the potential of fecal waste hydrolysates as a feedstock for the bioethanol production. However, the pre-treatment technologies are needed to implemented in order to extract cellulose and sugar resources (Taherzadeh and Karimi 2008; Sarkar et al. 2012). The previous research reports indicated the production of bioethanol by cattle and chicken manures through acid hydrolysis and enzymatic saccharification (Vancov et al. 2015; Yang et al. 2015).

Manures if not properly managed, present increased risk of pollution. If the fecal residues find way to water resources, they may increase chemical oxygen demand, reduce oxygen in water which in turn can cause the death of aquatic organisms (de Azevedo et al. 2017). The chemical composition of cattle manure depends on the diet and mainly comprised of undigested lignocellulose, which can be recovered. Approximately, 30–40% of the ingested food is utilized by the cattle (Konzen and Alvarenga 2005). Generally, manure is directed to lagooning, anaerobic digestion or combustion, which have environmental concerns in terms of CH_4, N_2O, and NH_3 emission (Gutiérrez et al. 2016). Alternatively, manure can be used as a source to produce bioethanol; however, manure is under-explored reservoir to produce bioethanol (Liao et al. 2014).

3.5.3 *Agriculture Crops*

A plenty of agricultural waste is produced across the globe that can be used as low-cost raw material to obtain valuable compounds (Akbar and Ali 2018). The agroindustry wastes are readily available to produce bioethanol such as cotton linters, vegetables and fruits wastes, coffee waste, wastepaper, etc. Although these wastes offer economic benefits, further innovations are required to optimize the production process and processing cost.

3.5.4 *Sugarcane Bagasse*

Bioethanol can be produced from sugarcane bagasse, but large-scale production has not yet become viable (Balat et al. 2008). Lignocellulosic materials are mainly comprised of hemicellulose, lignin, and cellulose (Saxena et al. 2009). Sugarcane bagasse can only be used as raw material for bioethanol production, if it is processed to yield fermentable sugars (Mosier et al. 2005).

The residue left after obtaining the sugarcane juice is termed as sugarcane bagasse. Brazil produces large amounts of sugarcane bagasse. In 2010–2011, an amount 208 million tons of sugarcane bagasse was attained by crushing 625 million tons of sugarcane (CONAB 2011). It is used for the production of electricity for commercial purposes, as well as by the ethanol and sugar mills as main source of energy (Rocha et al. 2012). Some of the applications for which the bagasse excess can be used include production of bioethanol, boards, animal feed, paper and pulp and furfural (Gálvez 2000; Mussatto et al. 2006).

Sugarcane bagasse is mainly produced by sugar- and alcohol-processing industries in Brazil, India, Cuba, China, Mexico, and Colombia (Martínez et al. 2003; Quintero et al. 2008; Hernández-Salas et al. 2009). Approximately, 1 ton of sugarcane is required to produce 280 kg of bagasse and 5.4×10^8 dry tons/year of sugarcane are processed across the globe (Cerqueira et al. 2007). Almost half of this agricultural waste is used in distillery facilities as a source of energy (Pandey et al. 2000) and remaining is stockpiled. Therefore, due to economic benefits the importance of bagasse as organic waste is gaining global interest in the production of biofuel (Adsul et al. 2004). Sugarcane bagasse being rich source of carbohydrates can be used for the production of ethanol by cellulose hydrolysis followed by fermentation (Botha and Blottnitz 2006).

The bagasse from sugarcane is comprised of cellulose (40–45%), hemicelluloses (30–35%), and lignin (20–30%) (Peng et al. 2009). The production of ethanol from sugarcane bagasse can be categorized into five major steps:

- Pre-treatment of biomass to remove non-sugar or non-fermentable moieties.
- Cellulose hydrolysis into simple sugars.
- Fermentation of sugars (hexoses).
- Separation and purification of ethanol.
- Effluent treatment.

Moreover, the process requires the detoxification and fermentation of pentoses produced from pre-treatment step. Ethanol production from sugarcane bagasse is presented in Fig. 3.3.

3.5.5 *Vegetables and Fruits*

Vegetable and fruit wastes are biodegradable in nature and produced in bulk amount, most of that is disposed on the land which can be environmental and health hazardous and can lead to the development of various diseases. In addition to post-harvest, losses due to poor storage facilities and improper processing and packaging generate considerable amounts of vegetable waste. Vegetable wastes can be categorized into rotten stock, shells, scraped portions of vegetables and peels (Singh et al. 2012).

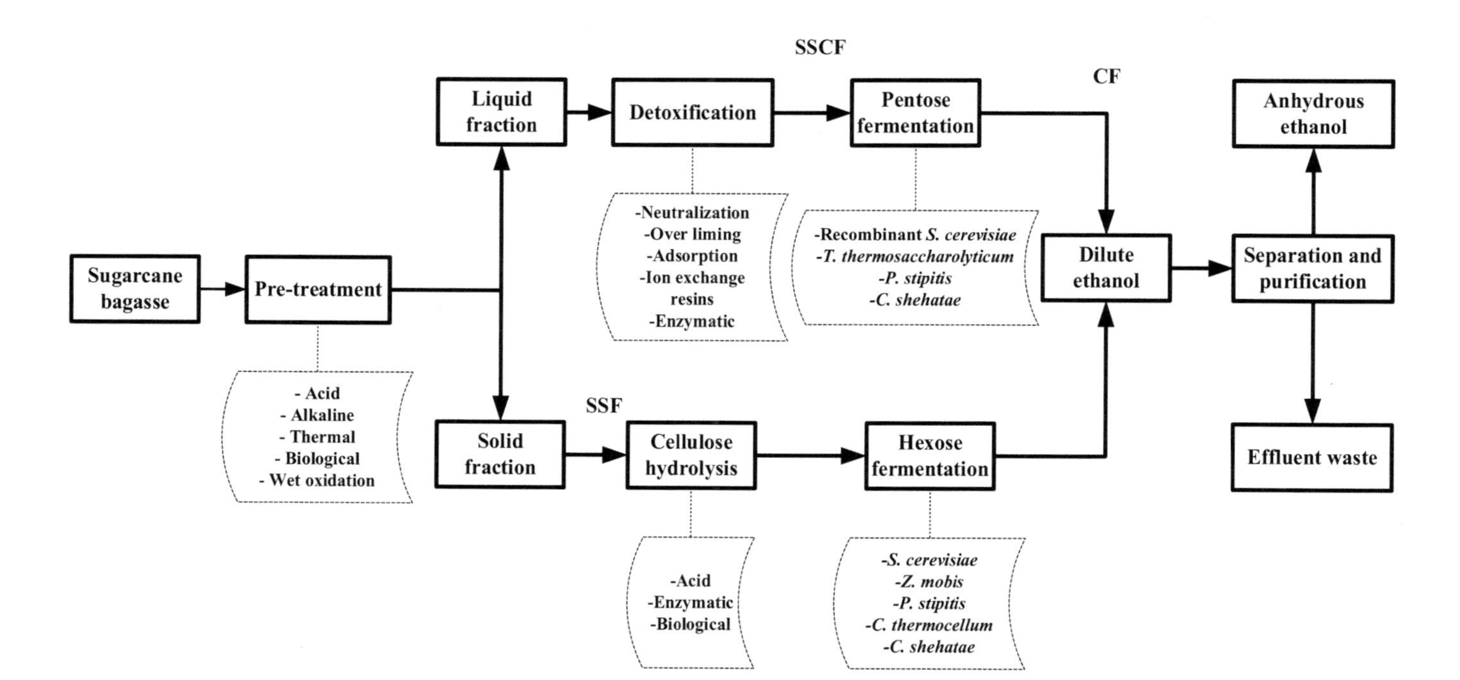

Fig. 3.3 Bioethanol production from sugarcane bagasse. (Adopted from Cardona et al. 2010). Here "CF = co-fermentation, SSF = simultaneous saccharification and fermentation, SSCF = simultaneous saccharification and co-fermentation." Whereas doted box connected to each step indicates the parameters required/applicable for the particular step

Vegetable and fruit wastes are organic in nature and rich in carbohydrates so can be used for production of biofuel. High humus content is generated by the natural decay of wastes. Research studies revealed that for bioethanol production, the carbohydrate-rich waste can be used as potential substrate. Carbohydrate-rich residues from vegetable- and fruit-processing industries can be used to prepare bioethanol by solid-state fermentation (Jørgensen et al. 2007; Laufenberg et al. 2003). The best suited substrate for production of bioethanol can be the agricultural waste as it has high cellulose and starch content, easily accessible and non-competitive with food (Tang et al. 2008).

The treatment of vegetable waste to produce biofuel can be categorized in three major steps:

- Pre-treatment,
- Saccharification and
- Fermentation.

Bioethanol, approximately equal to 442 billion liters, can be produced from ligno-cellulosic waste materials, while 491 billion liters of bioethanol can be yield annually from total crop residues and wastes, which is 16 times more than the actual world bioethanol production (Kim and Dale 2004).

3.5.6 *Organic Waste-Based Biomolecules for Bioethanol Production*

The global demand for biofuels is increasing and using food crops like sugarcane and corn interferes with the food security. Therefore, the biomolecules such as ligno-cellulose from agriculture waste can be a potential alternate raw material to produce bioethanol. The production of bioethanol from agricultural waste can be the solution to meet the fuel needs of next generation as it is abundant, sustainable, and cost-effective.

3.5.6.1 Cellulose

Cellulose-based bioethanol is viable and sustainable solution to fossil fuels and expected to be commercialized in future as transportation fuel (Kuhad et al. 2011; Wiesenthal et al. 2009). Three major components on which the cost of ethanol from cellulose biomass depends on include cellulase enzyme, feedstock, and non-enzymatic transformation (Humbird et al. 2011; Liu et al. 2016). The cellulose biomass has potential to produce 491 giga liters/year of bioethanol (Kim and Dale 2004). The previous research reports have revealed the sustainable conversion of cellulosic feedstock into bioethanol (Chen et al. 2010; Park and Kim 2012; Kim and Kim 2014). The pre-treatment of cellulosic biomass and ethanol yield efficiency are controlling factors for bioethanol production. The progressions have been made regarding the economic and technological assessments for feasibility of cellulosic biomass-based ethanol (Aden and Foust 2009). Approximately, 2 million metric tons/year of cellulosic feedstock are produced by Taiwan suitable to produce bioethanol (Council of Agriculture 2010). Taiwan government encouraged the enterprises and assists them in establishing biorefineries for agriculture waste processing. As a result, the production of cellulosic bioethanol has been increased by local companies in Taiwan (Wen et al. 2015).

3.5.6.2 Lignocellulose

Lignocelluloses are carbohydrate polymers and are essential constituent of the plant cell wall that are responsible for complexity of cell structure. The polymer matrix is comprised of cellulose (40–50% w/w), and a branched heteropolymer, hemicellulose (25–35% w/w), that can be utilized for the production of bioethanol. However, the hemicellulose moieties are covalently bound to a polymer called lignin (15–20% w/w) which makes carbohydrate depolymerization challenging (Rico et al. 2014). In lignocellulosic ethanol production, saccharification is the most important step where lignin-free polymers are depolymerized to monosaccharides using microbial xylanases and cellulases (Althuri et al. 2017).

Production of biofuel from lignocellulose biomass is gaining interest due to its non-competitive nature with human food demands, sustainability, and abundance (Ravindran and Jaiswal 2016). Conversion of lignocellulosic biomass to bioethanol has been indicated in the previous research reports (Cai et al. 2016). However, the pre-treatment step is essential to convert lignocellulosic biomass into simple sugars as bioethanol yield from microbial fermentation is mainly dependent on simple sugars (Alvira et al. 2010; Karimi and Taherzadeh 2016). Biofuel production at industrial scale is limited by the cost associated with hydrolytic enzymes (Xin and He 2013). Consolidated bioprocessing is best suited economical solution due to single step that involves enzymes production, cellulose saccharification, and microbial fermentation (Mahajan et al. 2014). Consolidated bioprocessing can be achieved by "recombinant cellulolytic strategy," in which microorganisms capable of producing high yield of ethanol are genetically engineered to hydrolyze cellulose directly or "native cellulolytic strategy" in which microorganisms are used to efficiently hydrolyze cellulose, but with low biofuel yield. Because of complex nature of cellulose degradation systems, "native cellulolytic strategy" would be more preferred (Jiang et al. 2017).

3.5.6.3 Pectin

Pectin is mainly composed of covalently linked galacturonic acids and may contain arabinose, rhamnose, xylose, and galactose (Mohnen 2008). 12–35% pectin is present in the cell wall of pectin-rich biomass (Mohnen 2008; Zhou et al. 2008). Whereas only 2–10% pectin is present in the cell walls of pectin non-rich biomass such as grasses, corn kernels, and woody material (Mohnen 2008). Pectin-rich biomass contains comparatively less lignin content than lignocellulosic biomass. This fact is substantial as lignin is non-fermentable to ethanol and hinders the enzymatic hydrolysis of hemicellulose and cellulose (Guo et al. 2009). Therefore, lignin needs to be separated from lignocellulose biomass before fermenting the carbohydrates. This separation requires extensive pre-treatments to degrade lignin that may convert some sugars into inhibitory molecules. The pectin-rich waste from fruit-processing industries is favorable biomass to produce bioethanol.

Pectin-rich biomass may need pre-treatment (to split the biopolymers) or removal of fermentation inhibitory substances such as removal of limonene in citrus waste (Grohmann et al. 1994; Wilkins et al. 2009). However, no further pre-treatment or removal of inhibitory compounds is required for fermentation of sugar beet pulp. Variety of ethanologens such as bacteria and yeast can be used to conduct fermentation of pectin-rich biomass. *E. coli* can ferment galacturonic acid, the main constituent of pectin. *S. cerevisiae* cannot naturally ferment galacturonic acid or pentose sugars but exhibits a homoethanol pathway. Broad range of pectin-rich materials such as pectin can be degraded by *Erwinia chrysanthemi.* The production of bioethanol from pectin-rich biomass has an additional benefit as waste material is already collected and partially pre-treated to ease enzymatic breakdown of the plant cell walls (Edwards and Doran-Peterson 2012).

3.6 Algal Feedstock for Bioethanol

Algae are the simple living organisms that bear chlorophyll and utilize sunlight for photosynthetic process. Algae can either be phototrophic or heterotrophic. Phototrophic algal species can use carbon dioxide in atmosphere and convert it into nutrients, for instance, carbohydrate. On the contrary, organic compounds as source are used by heterotrophic algae for their growth and development (Wen and Chen 2003).

Algal biomass is significantly important for the production of many kinds of biofuels including isobutene, diesel, hydrogen, and ethanol (Mussatto et al. 2010). Algal bioethanol production involves the fermentation of carbohydrates obtained from algae that include sugar, starch, and cellulose. The carbohydrates present in the algal structure have the capability to be used to produce ethanol following different hydrolytic processes. In case of aquatic algal species, their cells require no structural polymers like lignin and hemicellulose in water that are essential for terrestrial plants (John et al. 2011). This indicates the simplification of the bioethanol production process in algae.

Algal biomass is used to extract cellulose and starch by enzymatic hydrolysis or using mechanical shear, and thereafter they are used for the ethanol yields (John et al. 2011). Because of absence or negligible amount of lignin in algae, enzyme hydrolysis of algal cellulosic structure is rather simpler as compared with plant biomass. It has been reported that different algal species possess various biomass and starch contents after the process of oil extraction (John et al. 2011). Production of bioethanol from algal starch is identical to the conversion processes of sugars or starch to ethanol. Cellulosic biomass of both algae and plant shares the similar conversion techniques to ethanol. The other beneficial aspect of using algal biomass is no requirement of freshwater for their cultivation. Algal biomass can be cultivated using the wastewater from domestic and industrial sewage (Mussatto et al. 2010).

3.6.1 Microalgae for Bioethanol Production

Microalgae are prokaryotic or eukaryotic small organisms with the photosynthetic ability which may survive in extremely harsh environmental conditions with their unicelled and simple colony structures (Mata et al. 2010). The carbohydrate content of microalgae which mainly consist of starch can reach to 70 percent under particular conditions (Brányiková et al. 2011). They have very huge quantity of starch, like *Scenedesmus*, *Dunaliella*, *Chlorella*, *Chlamydomonas*, that are very beneficial for making of bioethanol. Majority of the microalgae including *Nannochloris sp.*, *Schizochytrium sp.*, *Botryococcus braunii, Chlorella sp.*,*Nitzschia sp.* contain at least 20% and up to 50% oil content (Chisti 2007).

Microalgal biomass can be transformed into bioethanol, biohydrogen, biodiesel, bio-oil, and biomethane through biochemical and thermochemical processes (Demirbas 2011). The utilization of microalgae as biofuel source is not something new in the world (Chisti 1980), but this is currently taking serious concern due to the increase in the petroleum cost and, most importantly, rising concern about the emission of GHGs in the environment by fossil fuel combustion (Sawayama et al. 1995).

Specific microalgae species possess the potential of producing larger amount of carbohydrates than lipids as reserved polymeric compounds. Such species are the best for producing bioethanol as carbohydrates could be isolated from microalgae to make sugars of fermentable nature. According to some estimates, microalgae can produce around 5000 to 15,000 gal of ethanol/acre/year (46,760–140,290 L/ha). This ethanol yield value from microalgae is many folds greater as compared with the yields from other feedstocks. *Chlorococum* sp. and Spirogira sp. are the blue-green algae which have been observed to deposit higher polysaccharide levels both as starch and in their complexed cell wall structure. This accumulation of starch in the microalgae species can be used for Bioethanol production (Harun et al. 2010; Eshaq et al. 2011). Harun et al. (2010) reported 60% greater ethanol yields from *Chlorococum sp.* for samples pre-extracted for lipids than those which remain as dry cells. It shows that microalgae could be utilized for the manufacturing of both ethanol- and lipid-based biofuels from the similar biomass to add to their economic worth (Jones and Mayfield 2012).

The microalgal species, mainly *Chlorella vulgaris,* has been regarded as a very good raw material for production of bioethanol owing to this fact that it has the potential of accumulating up to 37% of starch (dry weight). However, greater starch content could be also acquired for the best culture conditions (Hirano et al. 1997). Moreover, under various fermentation conditions, *Chlorococum* sp. were also utilized as a substrate to produce ethanol. Results indicated highest concentration of bioethanol, i.e., 3.83 g/L produced from 10 g/L of debris of lipid-extracted microalgae (Harun et al. 2010).

The production of ethanol from microalgae starts with the collection and drying of algal biomasses that are grown in an appropriate water condition. Grinding of the algal biomass is the next step performed followed by hydrolysis process and then the hydrolysed algal mass is fermented by adding microorganisms to produce ethanol and eventually the ethanol is separated and purified via distillation process (Demirbas 2010). The ethanol production utilizing microalgae as feedstocks could be performed in accordance with following steps.

The starch contained in microalgae is freed from the cells in the first step with the help of enzymes or mechanical processes. When the cells start to degenerate, fermentation process begins by the addition of yeast *S. cerevisiae* in the biomass. Bioethanol produced as a result of fermentation is then drained from the tank and pumped into a holding tank to be poured to distillation equipment. The production of ethanol was caused with process of photosynthesis by microalgae and anaerobic fermentation process (Hirano et al. 1997; Pimentel 2003; Pimentel and Patzek 2005; Demirbas 2011).

Macroalgae are plants that dwell in the marine environment and thus are usually found in coastal areas. Based on their pigment colors, macroalgae can be categorized as green seaweeds, red seaweeds, and brown seaweeds (Jung et al. 2013). Macroalgae contains lesser lipid content as compared to microalgae and thus can be distinguished based on lipid content and can also be differentiated from lignocellulosic materials due to the absence or low lignin concentration in their structures (Daroch et al. 2013).

Macroalgae possess lower lipid content and mainly have 35 to 74% carbohydrates and 5 to 3% proteins (Ito and Hori 1989). Macroalgae have different concentrations of carbohydrates, green algae contain 25 to 50%, red algae 30 to 60%, and brown algae have 30 to 50%. The species of macroalgae with the contents of highest polysaccharide include *Palmaria* (38 to 74%), *Porphyra* (40 to 76%), and *Ascophyllum* (42 to 70%). The algal species with higher levels of carbohydrates are represented in Table 3.1 (Hall and Payne 1997).

The bacterial species *Zymomonas mobilis* and the yeast *S. cerevisiae* are currently the most crucial microorganisms to produce bioethanol but these ethanol-producing microorganisms are very specific to the type of substrate. Horn et al. (2000a) have indicated the *Zymobacter palmae* had the potential to convert mannitol into ethanol in the seaweed extract in the presence of some oxygen supply. Moreover, the ethanol-producing ability of extracts of *Laminaria hyperborean* was also investigated, with its primary focus on the yeast *Pichia angophorae* and its capability of using laminarin and mannitol both as substrates. The early experiments with regard to extract of seaweed indicated that *Pichia angophorae* was capable of using both laminarin

Table 3.1 Carbohydrate contents (%) of various algal species

Algae	Carbohydrate content (%)
C. vulgaris	55.0
Chlamydomonas reinhardtii UTEX 90	60.0
Chlorococum sp.	32.5
S. obliquus CNW-N	51.8
Tetraselmis sp.CS-362	26.0
Ulva lactuca	55–60
Ascophyllum	42–70
Porphyra	40–76
Palmaria	38–74

and mannitol for the production of bioethanol (Horn et al. 2000b). Laminarin and mannitol present in the *L. hyperborean* extracts can be utilized to produce ethanol. *Z. palmae* could utilize mannitol for ethanol production, while *P. angophorae* was able to produce bioethanol from both carbohydrates.

Algae-based biofuels are significantly important owing to lower concentration of hemicellulose and lignin in algae (Harun et al. 2010). However, macroalgae have higher levels (at least 50%) of sugars which can be utilized in chemical conversion of bioethanol through fermentation process (Wi et al. 2009). However, the concentration of carbohydrates in red algae is affected by agar presence which is a polymer of galactopyranose and galactose. Recently, researchers seek to develop certain saccharification methods for higher yield of bioethanol during fermentation process by the release of galactose from agar and further unlocking of glucose from cellulose (Wi et al. 2009; Yoon et al. 2010).

3.7 Life-Cycle Analysis of Biofuel Production from Biowastes

The drastic increase in the human population has evolved the needs of efficient energy availability for the masses. This has impacted on general fossil fuel usage which are limited in quantity. There is a dire need for the renewable energy sources for the development of human beings (Ali et al. 2014). Biofuels come as an effective choice, generating sustainable energy through the biogeochemical cycles involving living sources. A well-managed biofuel production system is not only environment friendly but also cost-effective (Bano et al. 2019).

Life-cycle assessment (LCA) is an effective tool in environmental sciences to understand the very beginning of the process to right at its bottom end (Roy et al. 2012). For understanding the comprehensive environmental performance of a system

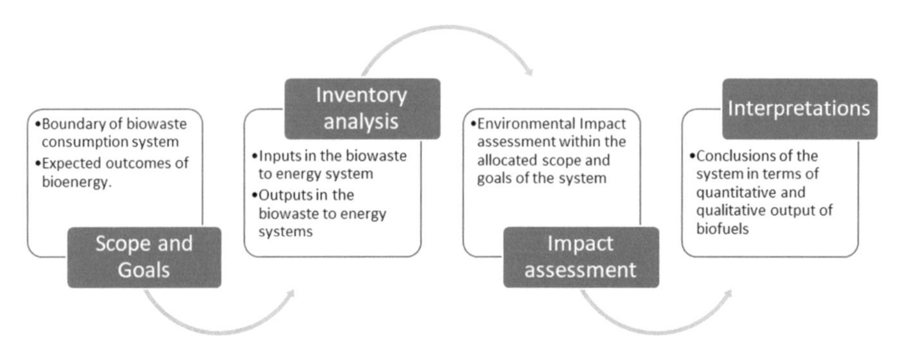

Fig. 3.4 Summary of methodology used in life-cycle analysis of biofuel production from biowastes

or product, the LCA is generally used, as it also performs the task of environmental impact assessment (EIA) (Papadaskalopoulou et al. 2019).

3.7.1 *LCA Methodology*

The concept of LCA in environmental sciences was evolved in 1960s, but the concept starts gaining the importance from 1990s, due to the increasing interest among the funders of research (Roy et al. 2012). For the evaluation of effective waste management, the LCA is an invaluable system, implemented with well-organized methodology (Pourreza et al. 2020).

As shown in Fig. 3.4, LCA methodology is comprised of four components, which are (a) setting scope and goals, (b) inventory analysis of life cycle, (c) impact assessment of life cycle, and (d) overall interpretation (Guerrero and Muñoz 2018).

3.7.2 *Scope and Goals*

The goals are set according to the boundaries assigned to the system. Boundaries are the limits of functional units such as space, quantity, and volumes of biowastes assigned in the system, while system is an expected output of the products in terms of biofuels or bioenergy from the system (Roy et al. 2012). Mostly for bioethanol production from biowaste, the well-to-wheel (WtW) approach is used for setting the scope and goals, where the use of raw materials for obtaining the final outputs is well considered (Morales et al. 2015).

3.7.3 Inventory Analysis of Life Cycle

This component includes the analysis of all inputs and outputs in the system set for the biofuel production from waste. The inputs can be the wastewater, agricultural waste, energy, microorganisms, etc. while the outputs can be the products such as bioethanol and its co-products such as biogas (like CO_2, CH_4, NH_4), water, fertilizers, etc. (Pourreza et al. 2020).

3.7.4 Impact Assessment of Life Cycle

The life-cycle impact assessments are based on assuming or recording the environmental impacts of the process of bioethanol formation from biowaste. These effects can be local or global such as eutrophication, acidification, emissions, ozone depletion, global warming, etc. (Roy et al. 2012).

3.7.5 Overall Interpretation

Interpretation of LCA is an assessment of the other three components discussed above, especially considering the overall inputs and outputs of the system and the environmental impact it may have. Similarly, the socio-economic benefits of the systems studied to produce bioethanol from biowastes are considered with an aim of sustainable use or replications of the models used in the process (Padi and Chimphango 2020).

3.7.6 Some Recent Studies of LCA of Biowaste Utilizations

In Gipuzkoa, Spain, two methods for waste management were comparatively analyzed through LCA. Their research concluded that both methods had different levels of output materials. The methods were production of energy in incinerator from mixed waste and recovery of materials after separation (Bueno et al. 2015). In Sakarya, Turkey, using LCA, Yay (2015) performed comparison of different waste management solutions, in which the best one was found having separation, landfilling, incineration and composting. Ripa et al. (2017) found through LCA that in area of Naples, USA, the absence of separation techniques is making the biowaste management difficult. Through LCA, in north and south of Europe, the environmental impact using biomethane of household for combined heat and power production systems was assessed (Evangelisti et al. 2017). In Italy, four different scenarios were studied for waste management using LCA. The best scenario included the separation of biowaste and production of biogas (Cremiato et al. 2018). In one study in China, it was found through LCA that incineration has better output than landfill, Zhou et al. (2018).

The reports on LCA of biowaste management are abundant but only some researchers have technically worked on conversion of biowaste into bioethanol (Papadaskalopoulou et al. 2019). Kalogo et al. (2007) compared the landfilling with the production of bioethanol from biowaste using LCA. From the organic component of municipal wastes of California, green house gas (GHG) emissions were assessed by LCA during bioethanol production (Chester and Martin 2009). Different waste management processes were comparatively analyzed in UK by using LCA, including bioethanol production from biowaste (Stichnothe and Azapagic 2009). Ebner et al. (2014) assessed the emissions of GHG associated with biowastes, including the production of bioethanol. Rizwan et al. (2018) developed a model through LCA in UAE, which has cost-effective recycling of biowaste for bioethanol production.

3.8 Sustainable Production and Utilization of Bioethanol (Future Perspective)

The advancement of living standards needs the quest for sustainable energy resources to meet the world's energy consumption demand (Demirbas 2010). There is dire need to find environment friendly, renewable and sustainable sources of energy by the combined efforts of government, industrial, and energy sectors (Shafiee and Topal 2009; Demirbas 2009). Liquid biofuels are prioritized for renewable energies as they represent 40% of total energy consumption in world and offer the lessening of greenhouse gas emissions (Demirbas 2009; Tan et al. 2008). Bioethanol is extensively used biofuel as an alternative in transportation since many decades. In 1984, Germany and France initiated the use of bioethanol in internal combustion engines (Demirbas and Karslioglu 2007). Until 1900s the bioethanol has been extensively used by US and Europe. The biofuel production has been depleted after World War II due to the cost associated with it till the oil crisis in the 1970s (Demirbas 2009). Since 1980s the use of bioethanol has gained interest, whereas in the current scenario it is used as an alternative source of fuel all around the globe.

The biofuel production policies need to be aligned with the United Nations' Sustainable Development Goals (SDGs) through prioritization in terms of long- and short-term goals. The factors that influence the implementation and prioritization include insufficient and unpredictable data across countries, lack of capacity and poor leadership, weak legal and policy frameworks, and land rights and ownership issues.

The extensive use of food crops to produce biofuels has been an important factor in contributing the high price of staple food since 2000 in the Brazil, Japan, US, and European Union. Although biofuel production offers significant opportunities in developing countries, at the same time, it presents threat to food security. There is a dire need to develop strong and well-established policies on feedstock farming, trading, and biofuel development as part of the United Nations' goals (UN 2015; Renzaho et al. 2017).

Acknowledgements The authors acknowledge the financial support provided by the University of Balochistan, Quetta through UBRF project for the completion of this study.

References

Aden A, Foust T (2009) Technoeconomic analysis of the dilute sulfuric acid and enzymatic hydrolysis process for the conversion of corn stover to ethanol. Cellulose 16(4):535–545

Adsul MG, Ghule JE, Singh R, Shaikh H, Bastawde KB, Gokhale DV, Varma AJ (2004) Polysaccharides from bagasse: applications in cellulase and xylanase production. Carbohyd Polym 57(1):67–72

Akbar A, Ali I (2018) Value-added by-products from sugar processing industries. In: Anal AK (ed.) Food processing by-products and their utilization, Wiley, p. 509–534

Akinsemolu AA (2018) The role of microorganisms in achieving the sustainable development goals. J Clean Prod 182:139–155

Alberts G, Ayuso M, Bauen A, Boshell F, Chudziak C, Peer Gebauer J et al (2016) International renewable energy agency (IRENA)

Ali I, Akbar A, Punnapayak H, Prasongsuk S, Yanwisetpakdee B (2014) Seven big challenges for Pakistan—and the lessons they could teach. Futurist 48:22–26

Althuri A, Gujjala LKS, Banerjee R (2017) Partially consolidated bioprocessing of mixed lignocellulosic feedstocks for ethanol production. Biores Technol 245:530–539

Alvira P, Tomás-Pejó E, Ballesteros MJ, Negro MJ (2010) Pretreatment technologies for an efficient bioethanol production process based on enzymatic hydrolysis: a review. Biores Technol 101(13):4851–4861

Arshad M, Khan ZM, Shah FA, Rajoka MI (2008) Optimization of process variables for minimization of byproduct formation during fermentation of blackstrap molasses to ethanol at industrial scale. Lett Appl Microbiol 47(5):410–414

Arshad M (2011) Bioethanol: A sustainable and environment friendly solution for Pakistan. Sci J COMSATS–Sci Vis 16–17

Arshad M, Zia MA, Asghar M, Bhatti H (2011) Improving bio-ethanol yield: using virginiamycin and sodium flouride at a Pakistani distillery. Afr J Biotechnol 10(53):11071

Arshad M, Ahmed S, Zia MA, Rajoka MI (2014) Kinetics and thermodynamics of ethanol production by *Saccharomyces cerevisiae* MLD10 using molasses. Appl Biochem Biotechnol 172(5):2455–2464

Arshad M, Hussain T, Iqbal M, Abbas M (2017) Enhanced ethanol production at commercial scale from molasses using high gravity technology by mutant *S. cerevisiae*. Braz J Microbiol 48(3):403–409

Arshad M, Zia MA, Shah FA, Ahmad M (2018) An overview of biofuel. Perspectives on water usage for biofuels production. Springer, Cham, pp 1–37

Arshad M, Bano I, Younus M, Khan A, Rahman A (2018b) Health concerns associated with biofuel production. In: Perspectives on water usage for biofuels production. Springer, Cham, pp 97–105

Balat M, Balat H, Öz C (2008) Progress in bioethanol processing. Prog Energy Combust Sci 34(5):551–573

Balat M, Balat H (2009) Recent trends in global production and utilization of bio-ethanol fuel. Appl Energy 86(11):2273–2282

Bano A, Xiaoming C, Prasongsuk S, Akbar A, Lotrakul P, Punnapayak H, Anwar M, Sajid S, Ali I (2019) Purification and characterization of cellulase from obligate Halophilic *Aspergillus flavus* (TISTR 3637) and Its prospects for bioethanol production. Appl Biochem Biotechnol 189:1327–1337. https://doi.org/10.1007/s12010-019-03086-y
Botha T, Von Blottnitz H (2006) A comparison of the environmental benefits of bagasse-derived electricity and fuel ethanol on a life-cycle basis. Energy Policy 34(17):2654–2661
Brányiková I, Maršálková B, Doucha J, Brányik T, Bišová K, Zachleder V, Vítová M (2011) Microalgae novel highly efficient starch producers. Biotechnol Bioeng 108(4):766–776
Bueno G, Latasa I, Lozano PJ (2015) Comparative LCA of two approaches with different emphasis on energy or material recovery for a municipal solid waste management system in Gipuzkoa. Renew Sustain Energy Rev 51:449–459
Cai D, Li P, Luo Z, Qin P, Chen C, Wang Y, Tan T (2016) Effect of dilute alkaline pretreatment on the conversion of different parts of corn stalk to fermentable sugars and its application in acetone–butanol–ethanol fermentation. Bioresour Technol 211:117–124
Cai J, He Y, Yu X, Banks SW, Yang Y, Zhang X, Bridgwater AV (2017) Review of physicochemical properties and analytical characterization of lignocellulosic biomass. Renew Sust Energy Rev 76:309–322
Cardona CA, Quintero JA, Paz IC (2010) Production of bioethanol from sugarcane bagasse: status and perspectives. Bioresour Technol 101(13):4754–4766
Cerqueira DA, Rodrigues Filho G, da Silva MC (2007) Optimization of sugarcane bagasse cellulose acetylation. Carbohyd Polym 69(3):579–582
Chastain J, Camberato J, Skewes P (2001) Confined animal manure managers certification program manual B poultry version. Poult Manure Prod Nutr, Content, p 2
Chastain JP, Camberato JJ (2004) Dairy manure production and nutrient content confined animal manure manager certification program manual dairy version, p 1–16
Chen F, Lu SM, Tseng KT, Lee SC, Wang E (2010) Assessment of renewable energy reserves in Taiwan. Renew Sust Energy Rev 14(9):2511–2528
Chen S, Wen Z, Liao W, Liu C, Kincaid RL, Harrison JH, Stevens DJ (2005) Studies into using manure in a biorefinery concept. Appl Biochem Biotechnol 124(1–3):999–1015
Chester M, Martin E (2009) Cellulosic ethanol from municipal solid waste: a case study of the economic, energy, and greenhouse gas impacts in California. Environ Sci Technol 43:5183–5189. https://doi.org/10.1021/es802788z
Chisti MY (1980) An unusual hydrocarbon. J Ramsay Soc 27–28:24–26
Chisti Y (2007) Biodiesel from microalgae. Biotechnol Adv 25(3):294–306
COA (2010) Agriculture statistics yearbook 2009. Council of Agriculture (COA)
CONAB (National Supply Company) (2011) The sugarcane production in Brazil will reach a new record. https://www.conab.gov.br. Accessed 12 Jan 2011
Cremiato R, Mastellone ML, Tagliaferri C, Zaccariello L, Lettieri P (2018) Environmental impact of municipal solid waste management using life cycle assessment: the effect of anaerobic digestion, materials recovery and secondary fuels production. Renew Energy 124:180–188
Cunha M, Romaní A, Carvalho M, Domingues L (2018) Boosting bioethanol production from Eucalyptus wood by whey incorporation. Bioresour Technol 250:256–264
Daroch M, Geng S, Wang G (2013) Recent advances in liquid biofuel production from algal feedstocks. Appl Energy 102:1371–1381
Davis R, Watts P, McGahan EJ (2012) Quantification of feedlot manure output for Beef-bal model upgrade. RIRDC
Demirbas A, Karslioglu S (2007) Biodiesel production facilities from vegetable oils and animal fats. Energy Sourc Part a 29(2):133–141
Demirbas A (2009) Biofuels securing the planet's future energy needs. Energy Convers Manage 50(9):2239–2249
Demirbas A (2010) Use of algae as biofuel sources. Energy Convers Manage 51(12):2738–2749
Demirbas MF (2011) Biofuels from algae for sustainable development. Appl Energy 88:3473–3480

De Azevedo A, Fornasier F, Da Silva SM, de Souza Schneider RDC, Hoeltz M, de Souza D (2017) Life cycle assessment of bioethanol production from cattle manure. J Clean Prod 162:1021–1030
Dias MO, Cunha MP, Jesus CD, Rocha GJ, Pradella JGC, Rossell CE, Bonomi A (2011) Second generation ethanol in Brazil: can it compete with electricity production? Biores Technol 102(19):8964–8971
Domínguez E, Romaní A, Domingues L, Garrote G (2017) Evaluation of strategies for second generation bioethanol production from fast growing biomass Paulownia within a biorefinery scheme. Appl Energy 187:777–789
Domínguez-Bocanegra AR, Torres-Muñoz JA, López RA (2015) Production of bioethanol from agro-industrial wastes. Fuel 149:85–89
Ebner J, Babbitt C, Winer M, Hilton B, Williamson A (2014) Life cycle greenhouse gas (GHG) impacts of a novel process for converting food waste to ethanol and coproducts. Appl Energy 130:86–93
Edwards MC, Doran-Peterson J (2012) Pectin-rich biomass as feedstock for fuel ethanol production. Appl Microbiol Biotechnol 95(3):565–575
Eshaq FS, Ali MN, Mohd MK (2011) Production of bioethanol from next generation feed-stock alga Spirogyra species. Int J Eng Sci Technol 3:1749–1755
Evangelisti S, Clift R, Tagliaferri C, Lettieri P (2017) A life cycle assessment of distributed energy production from organic waste: two case studies in Europe. Waste Manage 64:371–385
Gálvez LO (2000) Diversified productions in sugarcane agro-industry. In: Gálvez LO (ed) Handbook of sugarcane derivatives, 3rd edn. ICIDCA, Havana/Cuba, pp 3–17
González-García S, Mola-Yudego B, Murphy RJ (2013) Life cycle assessment of potential energy uses for short rotation willow biomass in Sweden. Int J Life Cycle Assess 18(4):783–795
González-García S, Moreira MT, Feijoo G (2012) Environmental aspects of eucalyptus based ethanol production and use. Sci Total Environ 438:1–8
Grohmann K, Baldwin EA, Buslig BS (1994) Production of ethanol from enzymatically hydrolyzed orange peel by the yeast *Saccharomyces cerevisiae*. Appl Biochem Biotechnol 45(1):315–327
Guerrero AB, Muñoz E (2018) Life cycle assessment of second generation ethanol derived from banana agricultural waste: environmental impacts and energy balance. J Clean Prod 174:710–717. https://doi.org/10.1016/j.jclepro.2017.10.298
Guo GL, Hsu DC, Chen WH, Chen WH, Hwang WS (2009) Characterization of enzymatic saccharification for acid-pretreated lignocellulosic materials with different lignin composition. Enzyme Microb Technol 45(2):80–87
Gutiérrez AS, Eras JJC, Billen P, Vandecasteele C (2016) Environmental assessment of pig production in Cienfuegos, Cuba: alternatives for manure management. J Clean Prod 112:2518–2528
Hall J, Payne G (1997) Factors controlling the growth of field populations of Hydrodictyon reticulatum in New Zealand. J Appl Phycol 9(3):229–236
Hansen DJ (2006) Manure as a nutrient source. In: Haering KC, Evanylo GK (eds). The Mid-Atlantic nutrient management handbook. MAWP 06-02, p 207
Harun R, Danquah MK, Forde GM (2010) Microalgal biomass as a fermentation feedstock for bioethanol production. J Chem Technol Biotechnol 85:199–203
Hernández-Salas JM, Villa-Ramírez MS, Veloz-Rendón JS, Rivera-Hernández KN, González-César RA, Plascencia-Espinosa MA, Trejo-Estrada SR (2009) Comparative hydrolysis and fermentation of sugarcane and agave bagasse. Bioresour Technol 100(3):1238–1245
Hill J, Nelson E, Tilman D, Polasky S, Tiffany D (2006) Environmental, economic, and energetic costs and benefits of biodiesel and ethanol biofuels. Proc Natl Acad Sci 103(30):11206–11210
Hirano A, Ueda R, Hirayama S (1997) CO_2 fixation and ethanol production with microalgal photosynthesis and intracellular anaerobic fermentation. Energy 22:137–142
Horn SJ, Aasen IM, Østgaard K (2000) Production of ethanol from mannitol by *Zymobacter palmae*. J Ind Microbiol Biotechnol 24:51–57
Horn SJ, Aasen IM, Østgaard K (2000b) Ethanol production from seaweed extract. J Ind Microbiol Biotechnol 25(5):249–254

Humbird D, Davis R, Tao L, Kinchin C, Hsu D, Aden A, Schoen P, Lukas J, Olthof B, Worley M, Sexton D (2011) Process design and economics for biochemical conversion of lignocellulosic biomass to ethanol: dilute-acid pretreatment and enzymatic hydrolysis of corn stover (No. NREL/TP-5100-47764). National Renewable Energy Lab. (NREL), Golden, CO (United States)

Ito K, Hori K (1989) Seaweed: chemical composition and potential food uses. Food Rev Int 5(1):101–144

Jiang Y, Xin F, Lu J, Dong W, Zhang W, Zhang M, Jiang M (2017) State of the art review of biofuels production from lignocellulose by thermophilic bacteria. Bioresour Technol 245:1498–1506

John RP, Anisha GS, Nampoothiri KM, Pandey A (2011) Micro and macroalgal biomass: a renewable source for bioethanol. Bioresour Technol 102(1):186–193

Jones CS, Mayfield SP (2012) Algae biofuels: versatility for the future of bioenergy. Curr Opin Biotechnol 23:346–351

Jørgensen H, Kristensen JB, Felby C (2007) Enzymatic conversion of lignocellulose into fermentable sugars: challenges and opportunities. Biofuels Bioprod Biorefin 1(2):119–134

Jung KA, Lim SR, Kim Y, Park JM (2013) Potentials of macroalgae as feedstocks for biorefinery. Bioresour Technol 135:182–190

Jung YH, Park HM, Kim DH, Park YC, Seo JH, Kim KH (2015) Combination of high solids loading pretreatment and ethanol fermentation of whole slurry of pretreated rice straw to obtain high ethanol titers and yields. Bioresour Technol 198:861–866

Kalogo Y, Habibi S, MacLean HL, Joshi SV (2007) Environmental implications of municipal solid waste-derived ethanol. Environ Sci Technol 41(1):35–41

Kargbo DM (2010) Biodiesel production from municipal sewage sludges. Energy Fuels 24(5):2791–2794

Karimi K, Taherzadeh MJ (2016) A critical review of analytical methods in pretreatment of lignocelluloses: Composition, imaging, and crystallinity. Bioresour Technol 200:1008–1018

Kim S, Dale BE (2004) Global potential bioethanol production from wasted crops and crop residues. Biomass Bioenergy 26(4):361–375

Kim S, Bae J, Choi O, Ju D, Lee J, Sung H, Um Y (2014) A pilot scale two-stage anaerobic digester treating food waste leachate (FWL): performance and microbial structure analysis using pyrosequencing. Process. Biochem 49(2):301–308

Kim TH, Kim TH (2014) Overview of technical barriers and implementation of cellulosic ethanol in the US. Energy 66:13–19

Konzen EA, Alvarenga RC (2005) Manejo e utilização de dejetos de suinos: aspectos agronômicos e ambientais. Embrapa Milho e Sorgo-Circular Técnica (INFOTECA-E). p 63

Kuhad RC, Gupta R, Khasa YP, Singh A, Zhang YHP (2011) Bioethanol production from pentose sugars: current status and future prospects. Renew Sustain Energy Rev 15(9):4950–4962

Laufenberg G, Kunz B, Nystroem M (2003) Transformation of vegetable waste into value added products: (A) the upgrading concept;(B) practical implementations. Bioresour Technol 87(2):167–198

Liao W, Liu Y, Hodge D (2014) Chapter 13-Integrated farm-based biorefinery. In: Vertes, NQBHA (ed.), Biorefineries. Elsevier, Amsterdam, p 255–270

Liao W, Liu Y, Liu C, Wen Z, Chen S (2006) Acid hydrolysis of fibers from dairy manure. Bioresour Technol 97(14):1687–1695

Liu H, Sun J, Leu S, Chen S (2016) Toward a fundamental understanding of cellulase-lignin interactions in the whole slurry enzymatic saccharification process. Biofuels Bioprod Biorefin 10:648–663

Mahajan C, Chadha BS, Nain L, Kaur A (2014) Evaluation of glycosyl hydrolases from thermophilic fungi for their potential in bioconversion of alkali and biologically treated *Parthenium hysterophorus* weed and rice straw into ethanol. Bioresour Technol 163:300–307

Martínez EA, Silva SS, Silva JBA, Solenzal AI, Felipe MG (2003) The influence of pH and dilution rate on continuous production of xylitol from sugarcane bagasse hemicellulosic hydrolysate by *C. guilliermondii*. Process Biochem 38(12):1677–1683

Mata TM, Martins AA, Caetano NS (2010) Microalgae for biodiesel production and other applications: a review. Renew Sust Energy Rev 14(1):217–232
McCollum DL, Echeverri LG, Busch S, Pachauri S, Parkinson S, Rogelj J, Riahi K (2018) Connecting the sustainable development goals by their energy inter-linkages. Environ Res Lett 13(3):033006
Miller JJ, Chanasyk DS, Curtis TW, Olson BM (2011) Phosphorus and nitrogen in runoff after phosphorus-or nitrogen-based manure applications. J Environ Qual 40(3):949
Mohnen D (2008) Pectin structure and biosynthesis. Curr Opin Plant Boil 11(3):266–277
Morales M, Quintero J, Conejeros R, Aroca G (2015) Life cycle assessment of lignocellulosic bioethanol: environmental impacts and energy balance. Renew Sustain Energy Rev 42:1349–1361. https://doi.org/10.1016/j.rser.2014.10.097
Mosier N, Wyman C, Dale B, Elander R, Lee YY, Holtzapple M, Ladisch M (2005) Features of promising technologies for pretreatment of lignocellulosic biomass. Bioresour Technol 96(6):673–686
Mussatto SI, Dragone G, Rocha GJ, Roberto IC (2006) Optimum operating conditions for brewer's spent grain soda pulping. Carbohyd Polym 64(1):22–28
Mussatto SI, Dragone G, Guimarães PM, Silva JPA, Carneiro LM, Roberto IC, Vicente A, Domingues L, Teixeira JA (2010) Technological trends, global market, and challenges of bio-ethanol production. Biotechnol Adv 28(6):817–830
Nasterlack T, Von Blottnitz H, Wynberg R (2014) Are biofuel concerns globally relevant? Prospects for a proposed pioneer bioethanol project in South Africa. Energy Sustain Dev 23:1–14
Nguyen TLT, Gheewala SH, Garivait S (2007) Energy balance and GHG-abatement cost of cassava utilization for fuel ethanol in Thailand. Energy Policy 35(9):4585–4596
Padi RK, Chimphango A (2020) Feasibility of commercial waste biorefineries for cassava starch industries: techno-economic assessment. Biores Technol 297:122461. https://doi.org/10.1016/j.biortech.2019.122461
Pandey A, Soccol CR, Nigam P, Soccol VT (2000) Biotechnological potential of agro-industrial residues. I: sugarcane bagasse. Bioresour Technol 74(1):69–80
Papadaskalopoulou C, Sotiropoulos A, Novacovic J, Barabouti E, Mai S, Malamis D, Kekos D, Loizidou M (2019) Comparative life cycle assessment of a waste to ethanol biorefinery system versus conventional waste management methods. Resour Conserv Recycl 149:130–139. https://doi.org/10.1016/j.resconrec.2019.05.006
Park YC, Kim JS (2012) Comparison of various alkaline pretreatment methods of lignocellulosic biomass. Energy 47(1):31–35
Paschalidou A, Tsatiris M, Kitikidou K (2016) Energy crops for biofuel production or for food?-SWOT analysis (case study: Greece). Renew Energy 93:636–647
Pejin D, Mojović LJ, Vučurović V, Pejin J, Denčić S, Rakin M (2009) Fermentation of wheat and triticale hydrolysates: a comparative study. Fuel 88(9):1625–1628
Peng F, Ren JL, Xu F, Bian J, Peng P, Sun RC (2009) Comparative study of hemicelluloses obtained by graded ethanol precipitation from sugarcane bagasse. J Agri Food Chem 57(14):6305–6317
Pimentel D (2003) Ethanol fuels: energy balance, economics, and environmental ımpacts are negative. Nat Resour Res 12:127–134
Pimentel D, Patzek TW (2005) Ethanol production using corn, switchgrass, and wood; biodiesel production using soybean and sunflower. Nat Resour Res 14:65–76
Pourreza MZ, Kabiri M, Ranjbar S, Joda F (2020) Multi-objective optimization of life cycle assessment of integrated waste management based on genetic algorithms: a case study of Tehran. J Clean Prod 247:119153. https://doi.org/10.1016/j.jclepro.2019.119153
IEA-ETSAP and IRENA (2013) Production of liquid biofuels technology brief
Quintero JA, Montoya MI, Sánchez OJ, Giraldo OH, Cardona CA (2008) Fuel ethanol production from sugarcane and corn: comparative analysis for a Colombian case. Energy 33(3):385–399
Ravindran R, Jaiswal AK (2016) A comprehensive review on pre-treatment strategy for lignocellulosic food industry waste: challenges and opportunities. Bioresour Technol 199:92–102

Razmovski R, Vučurović V (2012) Bioethanol production from sugar beet molasses and thick juice using *Saccharomyces cerevisiae* immobilized on maize stem ground tissue. Fuel 92(1):1–8
Renzaho AM, Kamara JK, Toole M (2017) Biofuel production and its impact on food security in low and middle income countries: implications for the post-2015 sustainable development goals. Renew Sustain Energy Rev 78:503–516
Rico A, Rencoret J, del Río JC, Martínez AT, Gutiérrez A (2014) Pretreatment with laccase and a phenolic mediator degrades lignin and enhances saccharification of Eucalyptus feedstock. Biotechnol Biofuels 7(1):6
Ripa M, Fiorentino G, Vacca V, Ulgiati S (2017) The relevance of site-specific data in Life Cycle Assessment (LCA). The case of the municipal solid waste management in the metropolitan city of Naples (Italy). J Clean Prod 142:445–460
Rizwan M, Saif Y, Almansoori A, Elkamel A (2018) Optimal processing route for the utilization and conversion of municipal solid waste into energy and valuable products. J Clean Prod 174:857–867
Rocha GJM, Gonçalves AR, Oliveira BR, Olivares EG, Rossell CEV (2012) Steam explosion pretreatment reproduction and alkaline delignification reactions performed on a pilot scale with sugarcane bagasse for bioethanol production. Ind Crop Prod 35(1):274–279
Roy P, Tokuyasu K, Orikasa T, Nakamura N, Shiina T (2012) A review of life cycle assessment (LCA) of bioethanol from Lignocellulosic biomass. Jpn Agric Res Q 46:41–57. https://doi.org/10.6090/jarq.46.41
Ruiz HA, Silva DP, Ruzene DS, Lima LF, Vicente AA, Teixeira JA (2012) Bioethanol production from hydrothermal pretreated wheat straw by a flocculating *Saccharomyces cerevisiae* strain–effect of process conditions. Fuel 95:528–536
Sadiq MB, Singh M, Anal AK (2018) Application of food by-products in medical and pharmaceutical industries. In: Anal AK (ed.) Food processing by-products and their utilization, Wiley, p 89
Saha BC, Nichols NN, Cotta MA (2011) Ethanol production from wheat straw by recombinant *Escherichia coli* strain FBR5 at high solid loading. Bioresour Technol 102(23):10892–10897
Sarkar N, Ghosh SK, Bannerjee S, Aikat K (2012) Bioethanol production from agricultural wastes: an overview. Renew Energy 37(1):19–27
Sawayama S, Inoue S, Dote Y, Yokoyama SY (1995) CO_2 fixation and oil production through microalga microalga. Energy Convers Manage 36:729–731
Saxena RC, Adhikari DK, Goyal HB (2009) Biomass-based energy fuel through biochemical routes: a review. Renew Sust Energ Rev 13(1):167–178
Schievano A, D'Imporzano G, Adani F (2009) Substituting energy crops with organic wastes and agro-industrial residues for biogas production. J Environ Manag 90(8):2537–2541
Shafiee S, Topal E (2009) When will fossil fuel reserves be diminished? Energy Policy 37(1):181–189
Singh A, Kuila A, Adak S, Bishai M, Banerjee R (2012) Utilization of vegetable wastes for bioenergy generation. Agric Res 1(3):213–222
Stafford W, Lotter A, Brent A, von Maltitz G (2017) United Nations university world institute for development economics research. Helsinki, Finland
Staniszewski M, Kujawski W, Lewandowska M (2007) Ethanol production from whey in bioreactor with co-immobilized enzyme and yeast cells followed by pervaporative recovery of product–Kinetic model predictions. J Food Eng 82(4):618–625
Steinfeld H, Gerber P, Wassenaar T, Castel V, de Haan C (2006) Livestock's long shadow: environmental issues and options. Food and Agriculture Organization of the United Nations, Rome, p 79–122
Stephen JL, Periyasamy B (2018) Innovative developments in biofuels production from organic waste materials: a review. Fuel 214:623–633
Stichnothe H, Azapagic A (2009) Bioethanol from waste: life cycle estimation of the greenhouse gas saving potential. Resour Conserv Recycl 53(11):624–630
Rajab S (2018) https://www.un.org/sustainabledevelopment/blog/2018/07/chanouf-farm-biofire/

Taherzadeh MJ, Karimi K (2008) Pretreatment of lignocellulosic wastes to improve ethanol and biogas production: a review. Int J Mol Sci 9(9):1621–1651
Taherzadeh MJ, Lennartsson PR, Teichert O, Nordholm H (2013) Bioethanol production processes. Biofuels production. Wiley, Hoboken, NJ, USA, pp 211–253
Tan KT, Lee KT, Mohamed AR (2008) Role of energy policy in renewable energy accomplishment: the case of second-generation bioethanol. Energy Policy 36(9):3360–3365
Tang YQ, Koike Y, Liu K, An MZ, Morimura S, Wu XL, Kida K (2008) Ethanol production from kitchen waste using the flocculating yeast *Saccharomyces cerevisiae* strain KF-7. Biomass Bioenergy 32(11):1037–1045
Uihlein A, Schebek L (2009) Environmental impacts of a lignocellulose feedstock biorefinery system: an assessment. Biomass Bioenergy 33(5):793–802
UN (2015) Agenda for sustainable development (A/RES/70/1), Transforming our world: the 2030. United Nations, New York
Vancov T, Schneider RCS, Palmer J, McIntosh S, Stuetz R (2015) Potential use of feedlot cattle manure for bioethanol production. Bioresour Technol 183:120–128
Vargas F, Domínguez E, Vila C, Rodríguez A, Garrote G (2015) Agricultural residue valorization using a hydrothermal process for second generation bioethanol and oligosaccharides production. Bioresour Technol 191:263–270
Venglovsky J, Sasakova N, Placha I (2009) Pathogens and antibiotic residues in animal manures and hygienic and ecological risks related to subsequent land application. Bioresour Technol 100(22):5386–5391
Vučurović DG, Dodić SN, Popov SD, Dodić JM, Grahovac JA (2012) Process model and economic analysis of ethanol production from sugar beet raw juice as part of the cleaner production concept. Bioresour Technol 104:367–372
Wen ZY, Chen F (2003) Heterotrophic production of eicosapentaenoic acid by microalgae. Biotechnol Adv 21(4):273–294
Wen PL, Lin JX, Lin SM, Feng CC, Ko FK (2015) Optimal production of cellulosic ethanol from Taiwan's agricultural waste. Energy 89:294–304
Wiesenthal T, Leduc G, Christidis P, Schade B, Pelkmans L, Govaerts L, Georgopoulos P (2009) Biofuel support policies in Europe: lessons learnt for the long way ahead. Renew Sustain Energy Rev 13(4):789–800
Wilkins MR, Widmer WW, Grohmann K (2009) x Simultaneous saccharification and fermentation of citrus peel waste by Saccharomyces cerevisiae to produce ethanol. Process Biochem 42(12):1614–1619
Wi SG, Kim HJ, Mahadevan SA, Yang DJ, Bae HJ (2009) The potential value of the seaweed Ceylon moss (*Gelidium amansii*) as an alternative bioenergy resource. Bioresour Technol 100:6658–6660
Xin F, He J (2013) Characterization of a thermostable xylanase from a newly isolated Kluyvera species and its application for biobutanol production. Bioresour Technol 135:309–315
Yang Q, Zhou S, Runge TM (2015) Dairy manure as a potential feedstock for cost-effective cellulosic bioethanol. Bioresour 11(1):1240–1254
Yay ASE (2015) Application of life cycle assessment (LCA) for municipal solid waste management: a case study of Sakarya. J Clean Prod 94:284–293
Yoon JJ, Kim YJ, Kim SH, Ryu HJ, Choi JY, Kim GS, Shin MK (2010) Production of polysaccharides and corresponding sugars from red seaweed. Adv Mater Res 93–94:463–466
Zhang M, Xie L, Yin Z, Khanal SK, Zhou Q (2016) Biorefinery approach for cassava-based industrial wastes: current status and opportunities. Bioresour Technol 215:50–62
Zhou W, Widmer W, Grohmann K (2008) Developments in ethanol production from citrus peel waste. In: Proceedings of the annual meeting of the Florida State Horticultural Society, vol 121
Zhou Z, Tang Y, Dong J, Chi Y, Ni M, Li N, Zhang Y (2018) Environmental performance evolution of municipal solid waste management by life cycle assessment in Hangzhou, China. J Environ Manage 227:23–33

Chapter 4
Marine Algae—Sustainable Raw Material for Bioethanol Production

Alina Roxana Lucaci, Dumitru Bulgariu, and Laura Bulgariu

Abstract The world is threatened by the diminishing and almost exhausting fossil fuels, climate change, and other negative effects of their use, which has led to the search for new, more environmental friendly energy sources that to successfully replace current requirements. Bioethanol has successfully proved that it fulfills the conditions to replace the fossil fuels, mainly because it is ecological and can be obtained from various renewable biomasses. In this context, the use of marine algae as raw material for the bioethanol production has several important advantages, such as its availability in large quantities in various regions of the world and requirement of only few harvesting and preparation steps, which means that their processing costs are low and are not considered food crops, which means that does not compete with the production of food. In this chapter, the utilization of marine algae as raw material for bioethanol production is analyzed to highlight the potential of this kind of biomass in the production of renewable energy. A summary characterization of marine algae, the main technological procedures currently used for the bioethanol production, and some economical considerations are summarized and discussed. On the basis of the aspects presented in this study, it can be said that marine algae is one of the most promising renewable sources which should be used as much as possible in the production of "green" energy.

Keywords Marine algae · Bioethanol production · Renewable energy · Environment protection

---

A. R. Lucaci · L. Bulgariu (✉)
Faculty of Chemical Engineering and Environmental Protection Cristofor Simionescu, Department of Environmental Engineering and Management, Technical University Gheorghe Asachi of Iaşi, Iaşi, Romania
e-mail: lbulg@tuiasi.ro

D. Bulgariu
Faculty of Geography and Geology, Department of Geology, Al.I.Cuza Univeristy of Iaşi, Iaşi, Romania

Filial of Iaşi, Romanian Academy, Branch of Geography, Iaşi, Romania

M. Arshad (ed.), *Sustainable Ethanol and Climate Change*,
https://doi.org/10.1007/978-3-030-59280-6_4

- The negative effects of fossil fuels have led to the search for new ecological and renewable energy resources.
- Bioethanol is the most widely used liquid biofuel for motor vehicles and this has determined the rapid growth of market.
- The use of marine algae as raw material for the bioethanol production has several important advantages.
- Before using marine algae biomass for large-scale ethanol production, several technological issues need to be addressed.

4.1 Introduction

The industrial activities and life quality of humans are highly dependent, in many countries of the World, on the fossil energy sources, such as coal, oil, or natural gas (Uihlein and Schbek 2009). Unfortunately, this fossil fuel dependence of the modern societies causes several important problems, including global warming, deterioration of air quality, oil splits, or acid rain (Borines et al. 2013), which have direct and severe consequences on the human health and the quality of the environment. For example, the transports are almost entirely dependent on fossil fuels, and used about 60% of the World fossil fuels and are responsible for the emission of 70% of carbon monoxide and 19% of CO_2 in atmosphere (Balat and Balat 2009). Furthermore, the experimental measurements have indicated that if in 2010 the level of CO_2 concentration in atmosphere was around 350–380 ppm, in 2020 this can attain a value of 450 ppm, if no action is taken (Kraan 2013). Another concern is related by the constant diminishing of fossil fuel resources. It is estimated that in Europe the oil resources will be exhausted in 40 years, the gas resources in 60 years, and the coal resources in 200 years (Balat 2010), if their consumption remains at the same level as now. In consequence, many countries will be forced to import fossil fuels, which will lead to a significant increase in sales prices.

All these important drawbacks related to the consumption of fossil fuels have determined the orientation of scientific research toward finding of new sources of clean energy, through the use of renewable resources in accordance with the principles of sustainable development, and this trend is also supported by the international political regulations. One way to reduce the consumption of fossil fuels is the use of biofuels obtained from renewable resources. It is widely accepted that the biofuels are an excellent alternative to traditional fossil fuels, mainly because they can be obtained from large available and renewable feedstock, as biomass, and their utilization generates lower or negligible levels of greenhouse gas and other pollutants (Dincer 2008; Balat and Balat 2009; Kraan 2013).

Of all biofuels, bioethanol has received a special attention, because it is the most widely used liquid biofuel for motor vehicles (Demirbas 2005) and for production of "green" energy, and these numerous utilizations have determined the rapid growth of market for this biofuel. Thus, if, in 2001, the world production of bioethanol was of 31 billion liters (Kim and Dale 2004), it is expected that in 2020 the production of this biofuel will exceed 150 billion liters. Under these circumstances, it is necessary to find a suitable raw material for the production of bioethanol, which must be efficient both economically and technologically, since it is not acceptable that the price of bioethanol to be higher than the price of conventional fuels.

In this moment, the raw biomass materials that can be used for the bioethanol production are divided into three categories, generic called "generations," namely: first generation—oleaginous biomass (corn, sugar cane, sugar beet, potato, etc.); second generation—lignino-cellulosic biomass (wood waste, lignino-cellulosic waste, etc.); and third generation—non-food crop biomass, such as marine algae biomass (Demirbas 2005; Ritslaid et al. 2010; Chen et al. 2015; Trivedi et al. 2015). In Table 4.1, the main characteristics of each generation of raw biomass used for bioethanol production are summarized. The selection of adequate biomass for the bioethanol manufacturing must take into account all these factors (see Table 4.1) so that the final price of the biofuel will be low (and thus accessible on market) and its production does not reduce the share of agricultural crops, because this will generate other significant problems such as the rise of vegetables cost and even starvation especially in developing countries (Pittman et al. 2011; Singh et al. 2011).

For example, USA and Brazil are the world's largest bioethanol producers, using corn, sugar cane, and potato as raw material, while sugar beet is the raw material for bioethanol extraction in Europe (Havlik et al. 2011). Unfortunately, the main disadvantage of using the first generation as raw materials for the extraction of bioethanol is the threat of food supplies that could affect the population due to the use of food resources (Arifin et al. 2014). As we are currently confronted with situations where people are malnourished or hungry, and the use of fuel generates higher food prices, the use of second-generation raw material for the bioethanol manufacturing can be a solution to this problem.

Table 4.1 The main characteristics of first, second, and third generation of biomass used for bioethanol production (Ritslaid et al. 2010; Chen et al. 2015; Trivedi et al. 2015)

	First generation	Second generation	Third generation
Food reserves	Food crops	Non-food crops	Marine algae
Use of land for cultivation	Agricultural land	Agricultural land	Marine water, lake water, wastewater
Conversion technologies	Sugar extraction, fermentation, distillation	Pretreatment, hydrolysis, fermentation, distillation	Hydrolysis, fermentation, distillation
Bioethanol yield	Low	Medium	High
Environmental impact	Low contribution to CO_2 mitigation	High contribution to CO_2 mitigation	High contribution to CO_2 mitigation
Advantages	Relatively light conversion processes	It does not compete with food resources	Very high growth rate
Disadvantages	Reduces the weight of food crops	Recalcitrant structure of raw material	Some technological difficulties

The second generation of raw material which can be used for the bioethanol production is lignino-cellulosic waste (such as waste wood processing, paper and pulp industry waste, agricultural waste, etc.) (Mohr and Raman 2010; Arifin et al. 2014). The second generation of raw material has some advantages but also disadvantages. The main advantage is that the raw materials used in the bioethanol manufacturing do not compete with the food reserves and are environmental friendly, while the disadvantages are related to the need to use advanced technology and costly conversion processes (Jambo et al. 2016).

The third generation is considered to be more beneficial. In this case, the raw material comes from the aquatic environment and more specifically focuses on the use of marine algae for the production of bioethanol. The main advantages of using marine algae as raw material are as follows:

I. Marine algae are considered as fast-growing plants that are available in large quantities in many regions of the World.
II. Marine algae do not need fertile land and freshwater for their cultivation (Demirbas and Karslioglu 2007), because they can be easily cultivated in different types of waters, and therefore they do not compete with the agricultural surface (Bixler and Porse 2011; Arifin et al. 2014; Jambo et al. 2016).
III. The cultivation of marine algae is favorable, they have a life cycle different from terrestrial plants, are more productive, can be collected at least five times a year, and do not need of chemical fertilizers (Goth and Lee 2010).
IV. Marine algae can convert solar energy into chemical energy with higher efficiency (6–8%) than terrestrial biomass (1.8–2.2%) (Jung et al. 2013), and for this reason are considered to have high CO_2 absorption properties.
V. Most of marine algae have high polysaccharides content, which makes ethanol extraction effective.

Although the use of marine algae for the bioethanol production has several important economical and environmental benefits, a major disadvantage has to be noted. Due to the particular composition of marine algae, which is different from most of terrestrial biomass, the common technologies must be modified and adapted, which involves some supplementary investments.

Therefore, the challenge in this moment is to find a suitable biomass which can be used as raw material for bioethanol production, which should be widely available and carbohydrate-rich. In addition, such biomass must be sustainable, should not use agricultural inputs (water, land, fertilizers, etc.), and most importantly not to be included in the human or animal food chain (Kraan 2013). A possible solution to this problem could be marine algae as this type of biomass meets most of the economical and technological requirements and its unique characteristics make this biomass the most promising potential sources for bioethanol production. Under these conditions, the cultivation and processing of marine algae are still current and important issues, for which solutions still have to be found, in order to improve the sustainability of the production of bioethanol.

In this chapter, the use of marine algae as potential raw material for bioethanol production is examined and discussed. Some basic information, which is useful for the marine algae characterization (classification, chemical composition, habitat environment, etc.), is reviewed. Also, the main technological procedures, which are currently used for the bioethanol production and some economical considerations are summarized and discussed. All the discussions summarized in this chapter support the idea that marine algae is one of the most promising renewable sources which should be used as much as possible in the production of "green" energy.

4.2 Brief Characterization of Marine Algae

Marine algae (usually called seaweeds) are pluri-cellular biological organisms whose size may vary in from a small microscopic size (3–10 μm) to large microscopic shapes (<70 m). From biological point of view, marine algae can be included in the category of inferior plants, mainly because they have some common elements with plants (stems and leaves), but unlike plants, the roots of marine algae are a thallus leaf. Due to this biological structure, the marine algae are photosynthetic organisms, which contain in their composition chlorophyll and other pigments, which print a color specific to the marine algae (Guiry 2012). During photosynthesis, the marine algae use CO_2 and sunlight as sources of energy, and produce most of the oxygen in the water required for all aquatic animals to breathe. This is the reason for considering marine algae as plants with high contribution to CO_2 mitigation (Chen et al. 2015), which can be used in the design of new ecological technological processes.

Up to now, about 30 thousand species of marine algae, grouped (as a function of their structure, nature of pigments, etc.) in 10–12 divisions (green algae, green–blue, red, brown, gold, etc.), are known (John et al. 2011; Kumari et al. 2013). Due to the large diversity, it is difficult to achieve a rigorous classification of marine algae according to scientific criteria. Most often, the classification of marine algae is based on the color of additional pigments, which even if not a rigorous scientific criterion divides the marine algae into three categories (Yu et al. 2002; Guiry 2012):

- Green algae (*Chlorophyta*)—their green color is mainly determined by the chlorophyll a and b;
- Red algae (*Rhodophyta*)—additional pigments which are responsible for the red-brick color of these algae are chlorophyll a, phycoerythrin, and phycocyanin;
- Brown algae (*Phaeophyta*)—principal photosynthetic pigments are chlorophyll a and c, b-carotene and other xanthophylls.

But besides the color, the marine algae have also different chemical constituents in their composition. Thus, the cell walls of green marine algae are mainly composed of cellulose and polysaccharides (such as mannose and xylose) that can be found in the structure of cell walls (Romera et al. 2007). In case of red marine algae, besides cellulose which represent the structural support of the cell walls, there may also be found different sulphated polysaccharides, like agar, xylose, mannitol, or carragenates (Romera et al. 2006, 2007). The brown marine algae have also a significant content of cellulose, alginate, and different sulphated polysaccharides such as mannitol, fucoidan, laminarin, etc. (Lodeiro et al. 2005; Romera et al. 2007).

As the performances of marine algae as a raw material in bioethanol production depend on their chemical composition, it is necessary to know which is their content in active components (such as polysaccharides), to assess whether the use of certain types of algae is economically feasible. In Table 4.2, the share of the main constituents that can be found in the structure of marine algae is summarized.

As can be seen from Table 4.2, the marine algae have relatively high content of polysaccharides, proteins, and lipids, and therefore they are suitable for use as raw material for the production of biofuels. But, because the content of polysaccharides, proteins, and lipids from marine algae varies in a large interval, depending on the category they belong to, it is expected that the green marine algae will be more suitable as raw material for the bioethanol production, while the brown marine algae are more appropriate in the biodiesel manufacturing processes.

However, in the utilization of marine algae as raw materials for the biofuels production, some particularities should be considered:

- Most of polysaccharides from marine algae composition (such as mannitol, alginate, xylose, fucoidan, etc.) (Guiry 2012) are different compared with those from structure of terrestrial plants.

Table 4.2 The weight of main constituents of some marine algae (Peterfi and Ionescu 1976; Renaud and Luong-Van 2006)

Type of marine algae	%, dry mass			
	Polysaccharide	Protein	Lipids	Ash
Green marine algae				
Ulva lactuca	29.5	13.9	1.8	16.6
Enteromorpha intestinalis	18.7	3.2	1.8	49.5
Caulerpa racemosa	14.7	6.9	4.4	47.7
Halimeda macroloba	2.7	4.6	2.5	64.4
Red marine algae				
Hypnea sp.	31.7	6.9	3.4	34.7
Portieria hornemannii	21.8	9.8	5.3	37.4
Gracilaria compressa	20.2	17.7	1.3	23.5
Laurencia majuscula	18.8	12.5	5.1	42.2
Brown marine algae				
Padina boryana	18.4	10.6	5.2	33.5
Hydroclathrus clathratus	18.3	4.2	2.9	49.4
Cystoseira barbata	17.4	13.5	1.9	20.4
Fucus virsoides	15.8	12.3	3.4	17.7
Dictyota ciliolata	15.2	4.1	7.8	47.2

- Compared to terrestrial plants, the marine algae have a low content of proteins and lipids (see Table 4.2), but a comparable content of starch, water, and minerals (Roesijadi et al. 2010);

• Carine algae has a very low content of lignin (Wegeberg and Felby 2010), and this is a real advantage in designing suitable technological facilities for the production of biofuels.

All these aspects should be taken into account in the selection of suitable marine algae as raw material for the bioethanol production, as they are strictly related to the technological and economic performance of the manufacturing process.

4.3 Growth and Preparation of Marine Algae Raw Material

Besides the advantages related to the chemical composition, other arguments in favor of using marine algae as raw materials for bioethanol are their high availability in many regions of the world and low cost of preparation.

It is well known that most of marine algae growths are on the water surface (up to 10 m on surface), but in the coastal areas of the seas and oceans they form stable multi-layer, perennial vegetation. In such coastal areas, the intensity of sun light and the water temperature in warm season are suitable for the growth of marine algae, and therefore annually large quantities of marine algae are obtained (Fig. 4.1a), without any costs. The marine algae which grow in such area are generally mechanically collected, dried in air, grinded, and can be used as raw material for bioethanol production. In addition, the collection of marine algae from beach zone is also in the benefit of tourism activities, because it helps to maintain the cleanliness of recreational areas.

Unfortunately, from technological point of view, the collection of marine algae, naturally grown in sea and ocean waters, has two major disadvantages, namely,

(a) (b)

Fig. 4.1 Naturally grown marine algae (**a**) and specialized farms for marine algae cultivation (**b**) (www.google.ro/search)

(i) The quantity of marine algae harvested strictly depends on the climatic conditions, and may vary within very wide limits depending on the season, which affects the continuity of the production process.
(ii) In the sea and ocean waters, more marine algae species can be found, and not all of them have the same efficiency in the production of bioethanol.

A solution that could minimize these disadvantages is the cultivation of marine algae in specialized farms using open-pound technology, and the naturally grown marine algae to be used additionally, when they meet the quality conditions required for the raw material. Thus, many farms specializing in the cultivation of marine algae (Fig. 4.1b), are located on the seashore or in special places, and are used to cultivate over 2500 species of seaweed worldwide. Due to their rapid development, farms specializing in the cultivation of marine algae provide a sufficient quantity of raw materials for the production of bioethanol on an industrial scale.

Whether the cultivation is done naturally or in specialized farms, after harvesting the marine algae requires only a few simple and inexpensive operations of preparation, which is also an important advantage which must be considered in the selection of a suitable raw material for the bioethanol production. In Fig. 4.2, the most important steps required for the preparation of marine algae before their utilization as raw material for the bioethanol production are illustrated.

All these operations can be performed in single or multiple steps, and have the role of transforming marine algae into an easy-to-handle material, which can be used later in the technological process. Under these conditions:

- ***marine algae harvesting*** is performed mechanically using special devices that collect the marine algae on the surface of the water;
- ***washing*** is done with demineralized water in order to remove the solid materials which were harvested with the marine algae, and the sea salts from marine algae leaves;

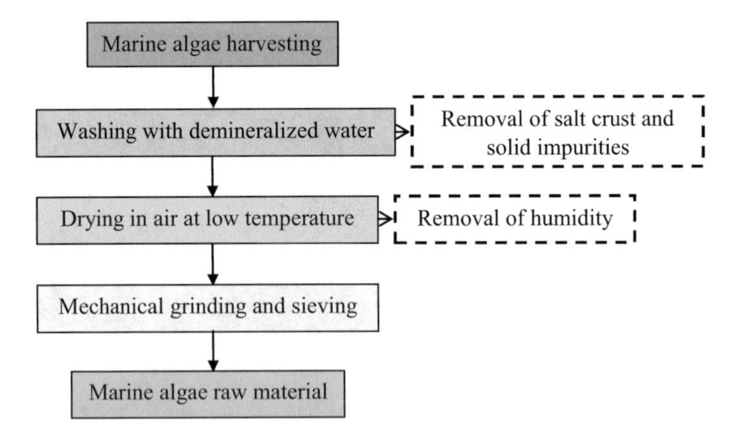

Fig. 4.2 Most important operations in the preparation of marine algae raw material

- ***drying*** is frequently done in hot air, at temperatures lower than 70 °C, and involves several steps until a complete drying of algae biomass;
- ***mechanically grinding and sieving***—dry marine algae biomass is comminuted until a suitable particle size (fine powder), by milling with specific sieve or by crushing.

The drying, mechanical grinding, and sieving operations are very important in the preparation of marine algae raw material, as the water removal and the obtaining small biomass particles (with high specific surface area) improve the efficiency of the other technological stages involved in the production of bioethanol. But since the last two operations require high energy consumption, their number should be carefully selected so as not to increase the cost of raw material preparation.

As can be seen from Fig. 4.2, the operations involved in the preparation of marine algae raw material are not significantly different from those required for ordinary terrestrial biomass, which means that the production of bioethanol has similar costs, in this respect. However, the finding of new energy-efficient alternatives that reduce the cost of marine algae preparation is still being studied extensively.

Considering all these aspects, it can be noted that the marine algae can be considered as suitable raw material for bioethanol production, if the following conditions are met: (i) the productivity of marine algae cultures is high, which involved reduced costs; (ii) available and can be purchased in sufficient quantities regardless of the season; (iii) the harvesting and preparation are easy and require simple and common operations; and (iv) their composition is adequate for the production of bioethanol (have a high content of polysaccharides).

4.4 Technological Stages of Bioethanol Production from Marine Algae Biomass

The use of marine algae as raw material for the production of bioethanol can be the solution to a number of important environmental issues such as reducing greenhouse gas emissions, controlling the pollution of natural or industrial water sources, or even saving agricultural land, which can be used for food crops.

From technological point of view, obtaining bioethanol from marine algae can be done generally by two processes: a biochemical conversion process which involves fermentation and a thermochemical conversion process, where the gasification of biomass occurs (Fig. 4.3). However, the higher content of water (80–90%) makes the marine algae to be more suitable for the biochemical conversion than for thermochemical conversion, to produce bioethanol (Ross et al. 2008).

Various biochemical conversion processes have been developed in order to transform the polysaccharides from marine algae into bioethanol, but in all cases the main important stages are pretreatment, hydrolysis (or saccharification), fermentation, and

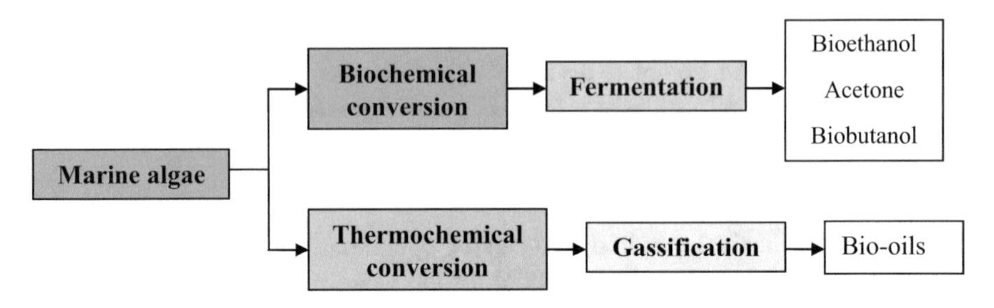

Fig. 4.3 Conversion processes used for bioethanol production from marine algae biomass

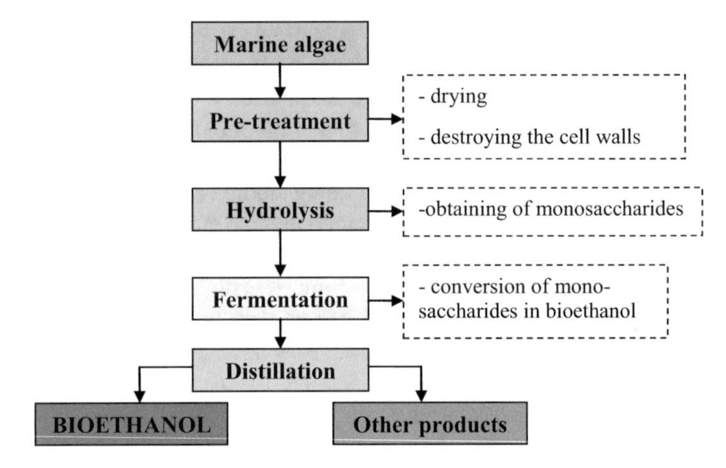

Fig. 4.4 Main stages of biochemical conversion of marine algae for bioethanol production

distillation. A schematic representation of a general biochemical conversion process is illustrated in Fig. 4.4.

4.4.1 Pretreatment

The pretreatment stage includes several operations which have as main result the drying of marine algae and the release of polysaccharides from marine algae cell walls. It is well known that the harvested algae contain almost 90% of water (Bibi et al. 2017). Therefore, reducing the water content of marine algae biomass is very important in order to obtain an easy-to-handle raw material. Most of the water content from marine algae is removed during the preparation steps (almost 50%), but this drying degree is not sufficient for further technological operations required for the bioethanol production.

Thus, various methods, such as low-pressure drying, sprig drying, freeze drying, and fluidized bed drying, can be used to reduce the water content of marine algae

Table 4.3 Type and content of polysaccharides in various marine algae (Chen et al. 2015; Sirajunnisa and Surendhiran 2016)

Class of marine algae	Type of polysaccharides	Marine algae	Content, % (w/w)
Green marine algae	Starch Cellulose	*Ulva fasciata*	43.0
		Caulerpa lentillifera	45.5
		Capsosiphon fulvescens	48.1
		Ulva lactuca	50.4
Red marine algae	κ-Carrageenan Cellulose	*Gracilaria verrucosa*	43.0
		Hypnea charoides	57.3
		Gelidium amansii	66.0
		Gracilaria sp.	76.7
Brown marine algae	Alginate Laminarin Mannitol	*Sargassum sp.*	41.6
		Laminaria japonica	51.5
		Ecklonia stolonifera	65.0
		Eisenia bicyclis	72.7

biomass in the range of 5–15% (Levine et al. 2010). However, each of these methods has advantages and disadvantages that need to be considered in designing an appropriate technological process.

But, only the drying of marine algae biomass is not enough to obtain a suitable raw material for the production of bioethanol. It is well known that most of polysaccharides from the cell walls of marine algae (see Table 4.3) are not freely available to hydrolyze, and for this reason, their release is necessary (Chen et al. 2015). This can be done by using steam pressure or ultrasonic methods and has the role of increasing the reaction zone and accessibility of polysaccharides for subsequent hydrolysis processes (Sirajunnisa and Surendhiran 2016). In this way, the cell walls of marine algae are destroyed by an additional supply of mechanical energy, facilitating the release of polysaccharides.

For example, Kim et al. (2015) have shown that by the pretreatment of red marine algae *Gelidium masii* with water steam at 121 °C and 60 min, the released polysaccharide content increased from 31.2 to 90.7%. Significant increase of the released polysaccharide content has been obtained and from other studies from literature where the hydrolysis of marine algae biomass was done by ultrasonic methods (Keris-Sen et al. 2014; Ward et al. 2014). Unfortunately, both of these methods required high energy consumption and multiple steps to achieve the desired results.

4.4.2 Hydrolysis

The release of the monosaccharides from the marine algae cell walls by hydrolysis is probably the most important stage in the industrial production of bioethanol using this

Table 4.4 The main advantages and disadvantages of chemical and enzymatic hydrolysis (Jambo et al. 2016)

Hydrolysis process	Advantages	Disadvantages
Chemical hydrolysis	Short time Reduced costs No products inhibition	Low yield High degree of corrosion High temperature
Enzymatic hydrolysis	High yield Low corrosion degree Low temperature	Long time High costs Products inhibition

raw material. During hydrolysis, the cell walls of marine algae will be depolymerized, and the polysaccharides will be transformed into free monomers, which can then be converted into bioethanol, by fermentation (Chandel et al. 2007; Vera et al. 2011).

From technological point of view, the hydrolysis of polysaccharides from marine algae structure can be done by (Abd-Rahim et al. 2014).

- chemical processes and
- enzymatic processes,

each with certain advantages and disadvantages (see Table 4.4).

4.4.2.1 Chemical Hydrolysis

The chemical hydrolysis supposed the treatment of marine algae biomass with mineral acids solution, in various experimental conditions. This hydrolysis method has been developed by the similarity with the method used for lignino-cellulosic biomass and aims to increase the fermentation efficiency. Among the mineral acids that may be used in the chemical hydrolysis, sulfuric acid is the most preferred one. Thus, the polysaccharides of the marine algae biomass can be efficiently hydrolyzed by their treatment with dilute H_2SO_4 solution and high temperature (El-Tayeb et al. 2012; Jang et al. 2012).

The H_2SO_4 acid solution has the role to break the hydrogen bond between polysaccharides chains, which leads to their transformation into an amorphous state. In such state, the amorphous polysaccharides can be easily hydrolyzed, due to the dissociation of glycosidic bonds, thereby producing monosaccharides (Hsu et al. 2011). In presence of H_2SO_4 solution and at high temperature, the process of chemical hydrolysis is efficient and requires a relatively short time. For example, the treatment of red marine algae, *Palmaria palmate*, with H_2SO_4 (0.04 mol/L) at 125 °C for 25 min, determines the significant increase of the hydrolysis yield for the carrageenan (main component) (Mutripah et al. 2014). Also, Khambhaty et al. (Khambhaty et al. 2012) have shown that the treatment of *Kappaphycus alvarezii* (red marine algae) with 0.9 mol/L H_2SO_4 solution at 100 °C allows to obtain a hydrolysis percent of 26.2%.

However, the disadvantages of the chemical hydrolysis processes summarized in Table 4.4 have determined the finding of other alternatives in a more efficient and less expensive way.

4.4.2.2 Enzymatic Hydrolysis

The enzymatic hydrolysis has been developed to overcome most of the disadvantages of the chemical hydrolysis. Many studies from literature (Demain et al. 2005; Balat 2010; Jambo et al. 2016) have indicated that the enzymatic hydrolysis has higher efficiency than the chemical hydrolysis (can achieve more than 80%), and these advantages (see Table 4.4) make this procedure to be more attractive for the production of bioethanol.

Generally, the enzymatic hydrolysis is based on the transformation of polysaccharides from marine algae biomass into monosaccharides, under the action of specific enzymes. The most used enzymes for this purpose are those from the *cellulases* category, which can be classified into three categories, namely, (Walker and Wilson 1991; Carere et al. 2008; Balat 2010; Yeh et al. 2010).

- *endo-glucanases*—acting on the inner part of the cellulose amorphous region, causing its hydrolysis;
- *exo-glucanases*—these hydrolyze cellulose by the splits of cellobiose units from the end of polymer chains;
- *β-glucosidase*acts on the formed cellobiases, and determines their splits into monomer units.

During enzymatic hydrolysis, the enzymes interact with marine algae (as substrate) at cellular level and a complex enzyme–substrate is formed. Then the enzymes will break down the cell walls of marine algae and thus the release of polysaccharides from their structure occurs. The released polysaccharides are then degraded to monosaccharides also by enzymes and these will be used in the fermentation stage to produce bioethanol (Jambo et al. 2016).

According to the studies from literature (Yeh et al. 2010; Meng and Ragauskas 2014; Jambo et al. 2016), the most important factor in achieving efficient enzymatic hydrolysis is the accessibility of hydrolysis enzymes to the surface of marine algae cell walls. This accessibility depends on several experimental factors, such as marine algae porosity, pH, time of hydrolysis, type of enzymes, pH of hydrolysis media, etc., and therefore the finding of the optimal conditions for the enzymatic hydrolysis in each case will be directly related to the efficiency of enzymatic hydrolysis process. For example, Kumar et al. (Kumar et al. 2013) have shown that the maximum enzymatic hydrolysis of Gracilaria verrucosa red algae is obtained by treating marine algae biomass with mixed enzymes cellulase and β-glucosidase, when after 36 h of incubation a concentration of 38.93 g/L sugar is obtained. In the same study, it has been mentioned that increasing hydrolysis time leads to a decrease in sugar yield, mainly due to the inhibition of hydrolysis enzymes. Also, the type of enzyme used in hydrolysis process influences its accessibility to the surface of marine algae cell walls, because different marine algae have different cell wall compositions and therefore have different enzyme specificities (Sirajunnisa and Surendhiran 2016). Consequently, optimal conditions for enzymatic hydrolysis should be experimentally determined in each case, and then applied extensively.

But, even if the experimental conditions are optimal, two more important drawbacks of enzymatic hydrolysis (see Table 4.5) should be solved in order to improve the efficiency and costs of hydrolysis process, namely, long time of hydrolysis and the impossibility of recovering enzymes at the end of the hydrolysis process. A solution for the long time of enzymatic hydrolysis could be the design of a hydrolysis process that takes place in two steps: first, the acid treatment of marine algae biomass (according to the chemical hydrolysis procedures), followed by the enzymatic hydrolysis, or vice versa. In this way, the main disadvantages of both hydrolysis methods are minimized, while the yield of sugar can increase significantly. Thus, Lee et al. (2013) have shown that the hydrolysis of raw Saccharina japonica by a two-step process determines the increase of the hydrolysis efficiency four times compared with single (acid or enzymatic) hydrolysis processes. However, the problem of recovering and recycling enzymes from the hydrolysis process still remains unresolved. A solution for this last aspect could be the immobilization of enzymes required for the hydrolysis process on magnetic nanoparticles.

It is well known that magnetic nanoparticles have numerous applications in industrial processes, because they are non-toxic, can be obtained by ecological processes, and have controlled particles sizes and high chemical reactivity (Wu et al. 2005; Johnson et al. 2009). In addition, the high efficiency of magnetic nanoparticles in the recovery of some catalysts (Govan and Gunko 2014) encourages their use for such applications as well.

Table 4.5 Microorganisms frequently used in the fermentation of hydrolyzed marine algae biomass in bioethanol production (Chen et al. 2015)

Class of marine algae	Type of monosaccharides	Microorganisms
Green algae	Cellulose	*Saccharomyces cerevisiae*
	Starch	
Red algae	Cellulose	*Saccharomyces cerevisiae*
	Carrageenan	
Brown algae	Alginate	*Escherichia coli*
	Laminarin	*Pichia angophorae* *Kluyveromyces marxianus* *Pachysolen tannophilus*
	Mannitol	*Zymobacter palmae* *Pichia angophorae* *Escherichia coli*

Table 4.6 The efficiency of bioethanol production using as raw material some types of marine algae

Marine algae	Hydrolysis	Fermentation	Bioethanol, g/g	Reference
Gracilaria salicornia	Enzymatic	*Escherichia coli*	0.079	Wang et al. 2011
Gracilaria verrucosa	Enzymatic	*Saccharomyces cerevisiae*	0.430	Kumar et al., 2013
Gelidium amansii	Acid	*Saccharomyces cerevisiae*	0.380	Park et al. 2012
Undaria pinnatifida	Acid	*Pichia angophorae*	0.330	Cho et al. 2013
Sargassum spp.	Enzymatic	*Saccharomyces cerevisiae*	0.170	Borines et al. 2013
Gracilaria sp.	Enzymatic	*Saccharomyces cerevisiae*	0.470	Wu et al. 2014
Laminaria japonica	Acid	*Escherichia coli*	0.410	Kim et al. 2011
Eucheuma cottonii	Enzymatic	*Saccharomyces cerevisiae*	0.025	Fakhrudin et al. 2014

In this approach, the enzymes to be used in the process of hydrolysis are immobilized on magnetic nanoparticle surfaces and mixed with the marine algae biomass. At the end of the hydrolysis process, the magnetic nanoparticle enzymes are separated using an extern magnetic device, and can be used in other hydrolysis cycles (Sharma et al. 2009; Chen et al. 2012). In this way, the recovery and recycling of hydrolysis enzymes can be achieved, and this possibility will directly influence the production costs (Sirajunnisa and Surendhiran 2016). However, replacing common chemical and physical methods for synthesis of magnetic nanoparticle with cleaner and cheaper biological methods is the new challenge in this area.

4.4.3 Fermentation

The monosaccharides (or simple "sugars") released after hydrolysis stage of marine algae can be transformed into bioethanol by fermentation in presence of suitable microorganisms. Various microorganisms, such as bacteria, fungi, or yeasts, are frequently used for the fermentation of hydrolyzed marine algae biomass, according to the well-known process:

$$\underset{\text{monosaccharide}}{C_6H_{12}O_6} \rightarrow \underset{\text{bioethanol}}{2\,C_2H_5OH} + 2\,CO_2 \tag{4.1}$$

Besides bioethanol important quantities of water and CO2 are also produced (Wei et al. 2013). Therefore, the selection of suitable microorganism in the fermentation stage is one of the main factors which influenced the efficiency of bioethanol production. The main criteria which are used for this purpose are selectivity of microorganisms, low accumulation of secondary products (water and CO2), efficiency and rate of fermentation process, as well as the cost of ease of handling. In Table 4.5, the most frequently used microorganisms for the fermentation of hydrolyzed marine algae biomass for the production of bioethanol are summarized.

The production of bioethanol from marine algae biomass can be done through two methods, namely, (Jelynne et al. 2014; Kim et al. 2015).

- ***separated hydrolysis and fermentation*** (SHF)—when the marine algae biomass is first hydrolyzed using a hydrolysis method mentioned in the previous paragraph, and then the hydrolyzed biomass is fermented, under the action of adequate microorganisms.
- ***simultaneous saccharification and fermentation*** (SSF)—when both the hydrolysis and the fermentation of marine algae biomass are carried out simultaneously, in a single step. In this case, the marine algae biomass, enzymes (required for hydrolysis), and microorganisms (for fermentation) are put together in a single reactor, and the monosaccharides formed during hydrolysis are immediately converted in bioethanol, by fermentation.

The main advantage of the separated hydrolysis and fermentation procedure (SHF) is related to the possibility of operating in optimal conditions and continuous mode, both for hydrolysis and fermentation processes, which is preferred at industrial scale (Jelynne et al. 2014). The disadvantages are determined by the inhibition of enzymes used during hydrolysis step, which cannot be reused any more, high operating time, and high cost of process operation (Balat 2011; Zhu et al. 2012).

Compared with SHF, simultaneous saccharification and fermentation (SSF) minimize the enzyme inhibition, the cost and time of operating processes, and simplify the technological flux (Gao et al. 2014; Liu et al. 2014). Since, during hydrolysis, formed monosaccharides are immediately converted into bioethanol by microorganisms, inhibition of enzymes is low, and also because the whole production process is carried out in a single reactor, the material cost is reduced. The main disadvantages of simultaneous saccharification and fermentation process are determined by the fact that not always both hydrolysis and fermentations processes can take place under optimal conditions, and that during fermentation the formed bioethanol can distorted the cells of the microorganisms, which reduces the fermentation efficiency (Liu et al. 2014).

Table 4.6 presents the efficiency of bioethanol production from some types of marine algae using the technological procedures described above.

Therefore, the finding of cheap and effective methods for the hydrolysis of polysaccharides from marine algae biomass composition, and efficient methods for the fermentation of various monosaccharides obtained after hydrolysis, is the most important aspect that still should be solved in order to improve the bioethanol production from marine algae biomass. In this context, more time and efforts are needed in order to generate a significant and reliable process for bioethanol production from this kind of raw material (Jambo et al. 2016).

4.4.4 Distillation

Distillation is the most common method that can be used to separate bioethanol from other fermentation compounds (Pacheco-Basulto et al. 2012). As is known, the distillation is a separation method based on component volatility and therefore experimental conditions must be rigorously controlled during operation. Most frequently, a distillation process requires the feedstock (the mixture resulted at the end of fermentation stage), which is introduced into a distillation reactor, where under the action of energy source (most often water stream), the separation of bioethanol occurs. As a function of the bioethanol quality requirements, additional purification steps may be added to obtain a high purity (Jambo et al 2016). The main disadvantages of distillation are the high energy consumption during separation process and long time of operation.

In the last years, the possibility of bioethanol separation by CO_2 supercritical extraction has received increased attention. This is because the CO_2 supercritical extraction method besides the low energy consumption, has also high efficiency and low operating time. In addition, most of the necessary CO_2 for supercritical extraction can be provided from fermentation stage, which is also in the benefit of environment.

4.5 Conclusions

Even if the commercial cultivation of marine algae has been started before 1950, only few companies are currently involved in the cultivation of marine algae for the bioethanol production. However, the increasing market requirements for bioethanol have recently led to a significant development of this production area. From an economic point of view, in the assessment of bioethanol production from marine algae biomass, the selection of the technological process must be appropriate and relevant, so that production costs can be minimized. Because raw material has the highest share in the cost of bioethanol production, intensive studies should focus on this aspect. In this context, the utilization of marine algae biomass as raw material for the bioethanol production has several important advantages, such as (i) they are available in large quantities in many regions of the world, (ii) they do not need fertile land and freshwater for their cultivation, (iii) they are more productive and do not need chemical fertilizers, and (iv) they have high CO_2 absorption properties, which increase their positive impact on environment, and have high polysaccharide content, which makes ethanol extraction effective.

However, in the utilization of marine algae biomass for the ethanol production at large scale, several important issues still need to be addressed. Thus, the productivity of marine algae cultures, the design of suitable cultivation system, cultivation conditions, and nutrients uptake, etc., are several important aspects that required considerable attention, before the marine algae are to be considered a feasible raw material for the production of bioethanol. All these have added the necessity to increase the economical viability of technological processes used to produce bioethanol, which involves the solving of technological particularities, the minimizing of energy consumption and production time, etc., so that the cost of bioethanol becomes competitive with those of fossil fuels. However, the numerous environmental advantages derived from the use of marine algae biomass as raw material for the production of bioethanol make all these efforts worthwhile, and this the considerable interest is highlighted by the high number of research studies in this field.

References

Abd-Rahim F, Wasoh H, Zakaria MR, Ariff A, Kapri R, Ramli N, Siew-Ling L (2014) Production of high yield sugars from Kappaphycus alvarezii using combined methods of chemical and enzymatic hydrolysis. Food Hydrocolloids 42:309–315

Arifin Y, Tanudjaja E, Dimyati A, Pinontoan R (2014) A second generation biofuel from cellulosic agricultural by product fermentation using *Clostridium* species for electricity generation. Energy Procedia 47:310–315

Balat H (2010) Prospects of biofuels for a sustainable energy future: a critical assessment. Energy Ed Sci Technol Part a 24:85–111

Balat M (2011) Production of bioethanol from lignocellulosic materials via the biochemical pathway: a review. Energy Convers Manag 52:858–875

Balat M, Balat H (2009) Recent trends in global production and utilization of bio-ethanol fuel. Appl Energy 86(11):2273–2282

Bibi R, Ahmad Z, Imran M, Hussaind S, Ditta A, Mahmood S, Khalid A (2017) Algal bioethanol production technology: a trend towards sustainable development. Renew Sustain Energy Rev 71:976–985

Bixler HJ, Porse H (2011) A decade of change in the seaweed hydrocolloids industry. J Appl Phycol 23:321–335

Borines MG, de Leon RL, Cuello JL (2013) Bioethanol production from the macroalgae Sargassum spp. Bioresour Technol 138:22–29

Carere CR, Sparling R, Cicek N, Levin DB (2008) Third generation biofuels via direct cellulose fermentation. Int J Mol Sci 9:1342–1360

Chandel AK, Chan E, Rudravaram R, Narasu ML, Rao LV, Ravindra P (2007) Economics and environmental impact of bioethanol production technologies: an appraisal. Biotechnol Mol Biol Rev 2:14–32

Chen H, Zhou D, Luo G, Zhang S, Chen J (2015) Macroalgae for biofuels production: process and perspectives. Renew Sustain Energy Rev 47:427–437

Chen WX, Yan Z, Yu WC, Xia WH (2012) Preparation and characterization of magnetic Fe3O4/CRGO nanocomposites for enzyme immobilization. Trans Nonferrous Met Soc China 22:162–168

Cho YK, Kim H, Kim SK (2013) Bioethanol production from brown seaweed, *Undaria pinnatifida* using NaCl acclimated yeast. Bioprocess Biosyst Eng 36:713–719

Demain AL, Newcomb M, Wu JHD (2005) Cellulase, clostridia, and ethanol. Microbiol Mol Biol Rev 69:124–154
Demirbas A (2005) Bioethanol from cellulosic materials: a renewable motor fuel from biomass. Energy Sourc 27:327–333
Demirbas A, Karslioglu S (2007) Biodiesel production facilities from vegetable oils and animal fats. Energy Sourc Part a 29:133–141
Dincer K (2008) Lower emissions from biodiesel combustion. Energy Sources Part a 30:963–968
El-Tayeb TS, Abdelhafez AA, Ali SH, Ramadan EM (2012) Effect of acid hydrolysis and fungal biotreatment on agro-industrial wastes for obtainment of free sugars for bioethanol production. Braz J Microbiol 43:1523–1535
Fakhrudin J, Setyaningsih D, Rahayuningsih M (2014) Bioethanol production from seaweed Eucheuma cottonii by neutralization and detoxification of acidic catalyzed hydrolysate. Int J Environ Sci Dev 5(5):455–458
Gao Y, Xu J, Yuan Z, Zhang Y, Liang C, Liu Y (2014) Ethanol production from high solids loading of alkali-pretreated sugar canebagasse with an SSF process. BioResources 9(2):3466–3479
Goth CS, Lee KT (2010) A visionary and conceptual macroalgae-based third-generation biothanol (TGB) biorafinery in Sabah, Malaysia as an underlay for renewble an sustainable development. Renew Sustain Energy Rev 14:842–848
Govan J, Gunko YK (2014) Recent advances in the application of magnetic nanoparticles as a support for homogeneous catalysts. Nanomaterials 4:222–241
Guiry MD (2012) The seaweed site: information on marine algae. https://www.seaweed.ie/algae/index.html.
Havlik P, Schneider OA, Schmid E, Bottcher H, Fritz S, Skalsky R, Aoki K, De Cara S, Kindermann G, Kraxner F, Leduc S, McCallum I, Mosnier A, Sauer T, Obersteiner M (2011) Global land-use implications of first and second generation biofuel targets. Energy Policy 39:5690–5702
Hsu CL, Chang KS, Lai MZ, Chang TC, Chang YH, Jang HD (2011) Pretreatment and hydrolysis of cellulosic agricultural wastes with a cellulase-producing Streptomyces for bioethanol production. Biomass Bioenergy 35:1878–1884
Jambo SA, Abdulla R, Azhar SHM, Marbawi H, Gansau JA, Ravindra P (2016) Areview on third generation bioethanol feedstock. Renew Sustain Energy Rev 65:756–769
Jang SS, Shirai Y, Uchida M, Wakisaka M (2012) Production of monosugar from acid hydrolysis of seaweed. Afr J Biotechnol 11:1953–1963
Jelynne P, Tamayo, Del Rosario E J (2014) Chemical analysis and utilization of *Sargassum* sp. As substrate for ethanol production. Iran J Energy Environ 5:202–208
John RP, Anisha GS, Nampoothiri KM, Pandey A (2011) Micro- and macro-algal biomass: a renewable source for bioethanol. Bioresour Technol 102(1):186–193
Johnson AK, Zawadzka AM, Deobald LA, Crawford RL, Paszczynski AJ (2009) Novel method for immobilization of enzymes to magnetic nanoparticles. J Nanopart Res 10:1009–1025
Jung KA, Lim SR, Kim Y, Park JM (2013) Potentials of macroalgae as feedstock for biorefinery. Bioresour Technol 135:182–190
Keris-Sen UD, Sen U, Soydemir G, Gurol MD (2014) An investigation of ultrasound effect on microalgal cell integrity and lipid extraction efficiency. Bioresour Technol 152:407–413
Khambhaty Y, Mody K, Gandhi M, Thampy S, Maiti P, Brahmbhatt H, Eswaran K, Ghosh PK (2012) Kappaphycus alvarezii as a source of bioethanol. Bioresour Technol 103:180–185
Kim DG, La HJ, Ahn CY, Park YH, Oh HM (2011) Harvest of *Scenedesmus* sp. with bioflocculant and reuse of culture medium for subsequent high-density cultures. Bioresour Technol 102:3163–3168
Kim HM, Wi SG, Jung S, Song Y, Bae HJ (2015) Efficencient approach for biethanol production fron red seaweed *Gelidium amasii*. Bioresour Technol 175:128–134
Kim S, Dale BE (2004) Global potential bioethanol production from wasted crops and crop residues. Biomass Bioenergy 26:361–375
Kraan S (2013) Mass cultivation of carbohydrate rich macroalgae, a possible solution for sustainable biofuel production. Mitig Adapt Strat Glob Change 18:27–46

Kumar S, Gupta R, Kumar G, Sahoo D, Kuhad RC (2013) Bioethanol production from *Gracilaria verrucosa*, a red alga, in a biorefinery approach. Bioresour Technol 135:150–156

Kumari P, Bijo AJ, Mantri VA, Reddy CRK, Jha B (2013) Fatty acid profiling of tropical marine macroalgae: an analysis from chemotaxonomic and nutritional perspectives. Phytochemistry 86:44–56

Lee JY, Li P, Lee J, Ryu HJ, Oh KK (2013) Ethanol production from Saccharina japonica using an optimized extremely low acid pretreatment followed by simultaneous saccharification and fermentation. Bioresour Technol 127:119–125

Levine RB, Pinnarat T, Savage PE (2010) Biodiesel production from wet algal biomass through in situ lipid hydrolysis and supercritical transesterification. Energy Fuels 24(9):5235–5243

Liu ZH, Qin L, Zhu JQ, Li BZ, Yuan YJ (2014) Simultaneou ssaccharification and fermentation of steam-exploded corn stover at high glucan loading and high temperature. Biotechnol Biofuel 7:167. https://doi.org/10.1186/s13068-014-0167-x

Lodeiro P, Cordero B, Barriada JL, Herrero R, Sastre de Vicente ME (2005) Biosorption of cadmium by biomass of brown marine macroalgae. Bioresour Technol 96:1796–1803

Meng X, Ragauskas AJ (2014) Recent advances in understanding the role of cellulose accessibility in enzymatic hydrolysis of lignocellulosic substrates. Curr Opin Biotechnol 27:150–158

Mohr A, Raman S (2010) Lessons from first generation biofuels and implications for the susteainability appraisal of second generation biofuels. Energy Policy 38:5892–5901

Mutripah S, Meinita MDN, Kang JY, Jeong GT, Susanto AB, Prabowo RE, Hong YK (2014) Bioethanol production from the hydrolysate of Palmaria palmate using sulfuric acid and fermentation with brewer's yeast. J Appl Phycol 26:687–693

Pacheco-Basulto JA, Hernandez-McConville D, Barrosa-Munoz FO, Hernandez S, Segovia-Hernandez JG, Castro-Montoya AJ, Bonilla-Petriciolet A (2012) Purification of bioethanol using extract batch distillation: simulation and experimental studies. Chem Eng Process: Process Intensif 61:30–35

Park JH, Hong JY, Jang HC, Oh SG, Kim SH, Yoon JJ, Kim YJ (2012) Use of *Gelidium amansii* as a promising resource for bioethanol: a practical approach for continuous dilute acid hydrolysis and fermentation. Bioresour Technol 108:83–88

Peterfi S, Ionescu A (1976) Algal handbook (in Romanian). Romanian Academy Publisher (Bucharest) 2:25–42

Pittman JK, Dean AP, Osundeko O (2011) The potential of sustainable algal biofuel production using wastewater resources. Bioresour Technol 102:17–25

Renaud SM, Luong-Van JT (2006) Seasonal variation in the chemical composition of tropical Australian marine macroalgae. J Appl Phycol 18:381–387

Ritslaid K, Kuut A, Olt J (2010) State of the art in bioethanol production. Agronomy Res 8(1):236–254

Roesijadi G, Jones SB, Snowden-Swan LJ, Zhu Y (2010) Macroalgae as a biomass feedstock: a preliminary analysis (PNNL-19944). https://www.pnl.gov/main/publications/external/technical_reports/PNNL-19944.pdf

Romera E, Gonzalez F, Ballester A, Blasquez ML, Munoz JA (2006) Biosorption with algae: a statistical review. Critical Rev Biotechnol 26:223–235

Romera E, Gonzalez F, Ballester A, Blazquez ML, Munoz JA (2007) Comparative study of biosorption of heavy metals using different types of algae. Bioresour Technol 98:3344–3353

Ross A, Jones JM, Kubacki ML, Bridgeman TG (2008) Classification of macroalgae as fuel and its thermochemical behaviour. Bioresour Technol 99:6494–6504

Sharma YC, Srivastava V, Singh VK, Kaul SN, Weng CH (2009) Nano-adsorbents for the removal of metallic pollutants from water and wastewater. Environ Technol 30:583–609

Singh A, Nigam PS, Murphy JD (2011) Renewable fuels from algae: an answer to debatable land based fuels. Bioresour Technol 102:10–16

Sirajunnisa AR, Surendhiran D (2016) Algae- a quintessential and positive resource of bioethanol production: a comprehensive review. Renew Sustain Energy Rev 66:248–267

Trivedi J, Aila M, Bangwal DO, Kaul S, Garg MO (2015) Algae-based biorefinery-how to make sense. Renew Sustain Energy Rev 47:295–307

Uihlein A, Schbek L (2009) Environmental impacts of a lignocellulosic feedstock biorefinery system: an assessment. Biomass Bioenergy 33:793–802

Vera J, Castro J, Gonzales A, Moenne A (2011) Seaweed polysaccharides and derived oligosaccharides stimulate defense responses and protection against pathogens in plants. Mar Drugs 9:2514–2525

Walker LP, Wilson DB (1991) Enzymatic hydrolysis of cellulose: an overview. Bioresour Technol 36:3–14

Wang X, Liu X, Wang G (2011) Two-stage hydrolysis of invasive algal feedstock for ethanol fermentation. J Integr Plant Biol 53(3):246–252

Ward AJ, Lewis DM, Green FB (2014) Anaerobic digestion of algae biomass: a review. Algal Res 5:204–2014

Wegeberg S, Felby C (2010) Algae biomass for bioenergy in Denmark: biological/technical challenges and opportunities. https://www.bio4bio.dk/~/media/Bio4bio/publications/Review_of_algae_biomass_for_energy_SW_CF_April2010.ashx.

Wei N, Quarterman J, Yong SJ (2013) Marine macroalgae: an untapped resource for producing fuels and chemicals. Trends Biotechnol 31:1–8

Wu FC, Wu JY, Liao YJ, Wang MY, Shih IL (2014) Sequential acid and enzymatic hydrolysis in situ and bioethanol production from *Gracilaria* biomass. Bioresour Technol 156:123–131

Wu LL, Yuan XY, Sheng J (2005) Immobilization of cellulose in nanofibrous PVA membranes by electrospinning. J Membr Sci 250:167–173

www.google.ro/search

Yeh AI, Huang YC, Chen SH (2010) Effect of particle size on the rate of enzymatic hydrolysis of cellulose. Carbohydr Polym 79:192–199

Yu S, Blennow A, Bojko M, Madsen F, Olsen CE, Engelsen SB (2002) Physico-chemical characterization of floridean starch of red algae. Starch 54:66–74

Zhu M, Li P, Gong X, Wang J (2012) A comparison of the production of ethanol between simultaneous saccharification and fermentation and separate hydrolysis and fermentation using unpretreated Cassava pulp and enzyme cocktail. Biosci Biotechnol Biochem 76(4):671–678

Chapter 5
Lignocellulosic Biomass and Microbial Genome Engineering for Sustainable Ethanol Production: An Overview

Muhammad Noman, Muhammad Shahid, Temoor Ahmed, Muhammad Rizwan Javed, Natasha Manzoor, Irfan Mazoor, Asad Ali Shah, Awais Maqsood, and Muhammad Arshad

Abstract In the last few decades, various scientific and technological developments in the fermentation process have significantly contributed to the progress of the ethanol industry worldwide. These major contributions have expanded our view about fermentation in first- and second-generation ethanol production. Currently, advanced technologies are available to produce ethanol from lignocellulosic biomass (LCB) such as sugarcane, sugar beet, corn, and other feedstocks. The LCB is composed of various 5- (e.g., arabinose and xylose) and 6- (e.g., glucose, galactose, and mannose) carbon sugars. The efficiency of ethanol processing needs to be enhanced further to meet the Sustainable Development Goals (SDGs) of the United Nations. Conventional genome modification techniques like mutation and selection are being extensively employed to over-produce ethanol through microbial fermentation. In addition to conventional techniques, this chapter evaluates the potential application of clustered regularly interspaced short palindromic repeats (CRISPR) associated genome engineering of potential microbes for enhanced ethanol efficiency. The CRISPR/Cas9 is a groundbreaking technology and has engineered microbial genomes for desirable traits in an efficient and advanced way. Genome alterations of various microbes including bacteria, fungi, algae, and *Saccharomyces cerevisiae* have been accomplished through this technique for various applications. In this chapter, we summarized the major discoveries of CRISPR/Cas9-mediated genome editing in microorganisms and deliberate how these discoveries can be utilized for the sustainability and enhancement of ethanol production.

M. Noman · M. Shahid (✉) · T. Ahmed · M. R. Javed · I. Mazoor · A. A. Shah · A. Maqsood
Department of Bioinformatics and Biotechnology, Government College University, Faisalabad 38000, Pakistan
e-mail: mshahid@gcuf.edu.pk

N. Manzoor
Institute of Soil and Environmental Sciences, University Agriculture Faisalabad, Faisalabad, Pakistan

M. Arshad
Jhang-Campus, University of Veterinary and Animal Sciences, Lahore, Pakistan

M. Arshad (ed.), *Sustainable Ethanol and Climate Change*,
https://doi.org/10.1007/978-3-030-59280-6_5

Keywords Ethanol · Efficiency improvement · CRISPR-Cas9 · Genome editing · Microbes

Due to the growing consumer demands, the efficiency of ethanol processing needs to be enhanced. The conventional feedstocks and genome engineering techniques are not sufficient to meet the consumer demands for ethanol. Hence, the CRISPR/Cas9 is a groundbreaking technology and can be employed to manipulate microbial cells for enhanced ethanol production. This chapter summarized the efforts and discoveries related to CRISPR/Cas9 microbial genome engineering and the prospects of utilizing this technology for sustainability and enhancement of ethanol production. The use of efficient lignocellulosic biomass and CRISPR/Cas9-mediated genome engineering of microbes will ultimately help to achieve the United Nations sustainable development goal "*Ensure access to affordable, reliable, sustainable and modern energy for all*".

5.1 Introduction

During the twenty-first century, one of the important challenges is to meet the increasing requirements of energy for industrial procedures as well as for heating and transportation purposes (Saxena et al. 2015). In light of increasing energy crisis, there is an utmost need to develop eco-friendly fuels that would replace pollution-bearing fossil fuels. A biofuel is an ideal option that has emerged to meet the requirements in a viable way. Biofuels are attractive due to their sustainability, low carbon emission (reducing global warming), easy degradability, etc. (IPCC 2018; Ghosh et al. 2019). Several factors (Fig. 5.1) motivated the stakeholders to encourage the use of biofuels because of decreasing oil reserves, increasing prices of petroleum goods and their rising environmental and health issues (Chang et al. 2017; Arshad 2010). During recent years, the usage and production of biofuels have been increasing dramatically. From 2000 to 2007, the consumption and production rise from 18.2 to 60.6 billion liters, approximately 85% of which is bioethanol. Bioethanol is practically valuable and one of the very common biofuels that can be produced from a variety of inexpensive substrates (Fulton et al. 2004; Arshad et al. 2017). According to an estimate, it can decrease the emission of greenhouse gas by almost 30–85% as compared to gasoline. Currently, the USA and Brazil are the largest producers of bioethanol, producing 49.6 and 38.3% of global production, respectively. There is a growing interest in bioethanol production worldwide, which is evident by the production of 85 billion liters in 2011 (Coyle 2007; Fulton et al. 2004).

Numerous groups of structurally different raw materials can be processed into bioethanol by different thermochemical or biochemical methods. Thermochemical processes of bioethanol production involve the pyrolysis or gasification of biomass, whereas biochemical processes include the hydrolysis of the biomass into fermentable monomeric sugar followed by microbe-mediated fermentation into ethanol (Bhatia et al. 2017; Arshad et al. 2018). Lignocellulose is considered to be the most abundant feedstock for bioethanol production (Zhou et al. 2017). The biomass can be generated by utilizing the farmlands since the whole plant can be used as feedstock. Moreover, the economically efficient feedstock can be developed by cultivating different a variety of crops demanding less fertilizer treatment on the same farmland. However, the use of lignocellulosic biomass (LCB) as a raw material is limited due to its complex and recalcitrant nature (Zabed et al. 2016).

Chemically, LCB encompasses lignin, hemicellulose, and the cellulose (Bhatia et al. 2016). Inside LCB, cellulose is covered by hemicelluloses thus forming a complex. The resulting complex obstructs the activity of cellulase enzyme by restricting the entry of the enzyme into the complex under normal conditions. The cellulose-hemicellulose complex is further surrounded by lignin which inhibits the action of hydrolytic enzymes by chelating the enzymes. Therefore, the pretreatment of LCB is necessary in order to open the chemical structure for cellulase enzyme to work efficiently (Kumar et al. 2017). Pretreatment processes along with fermentable sugars also produce various toxic by-products such as furan aldehyde, organic acids, and phenolic compounds that affect the microbial growth during fermentation. Therefore, toxic inhibitory by-products can be removed through detoxification methods which increases the process cost (Xu et al. 2015). *Saccharomyces* spp., *Kluyveromyces* spp., and *Candida* spp. are the efficient microbes for the production of bioethanol from LCB (Bhatia et al. 2017). Additionally, different strategies including genome engineering can be used to enhance substrate utilization, improve tolerance to inhibitory end products, and construction of novel biosynthetic pathways to improve bioethanol production.

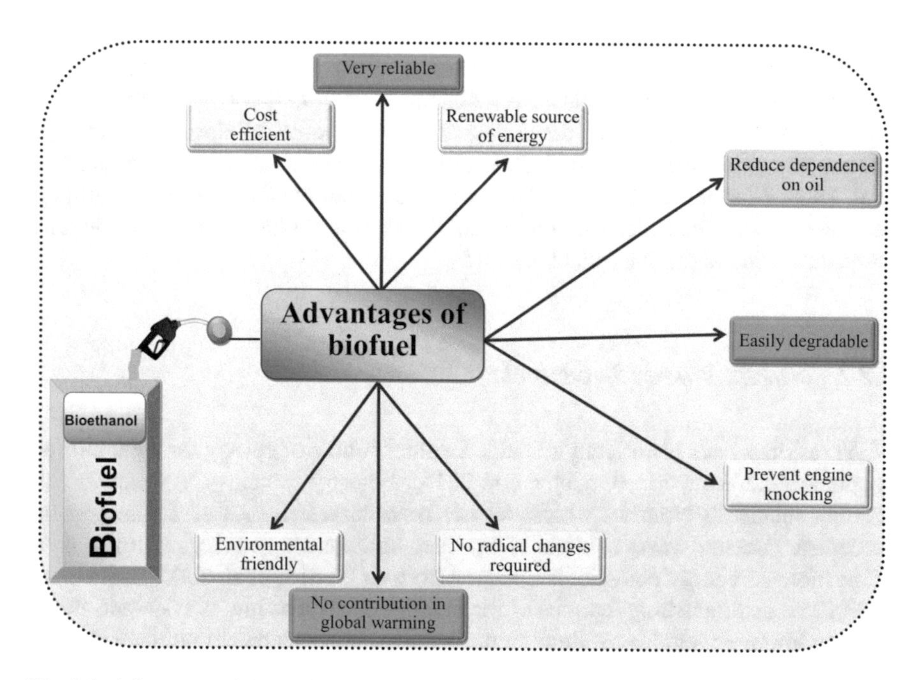

Fig. 5.1 Advantages of the biofuels over fossil fuels

To modify the activities and function of microorganisms, genetic transformation method is used through which heterologous gene expression is achieved. For this purpose, genome engineering is the way of producing some specific variations in the innate genome to change the function and activity of the microorganisms. It comprises the gene deletion, addition, and disruption at a specific site to modify the efficiency of microorganisms (Garneau et al. 2010; Wiedenheft et al. 2012). The efficiency of the microorganisms for a specific industrial process is also enhanced through specific or random mutation. Clustered regularly interspaced short palindromic repeats (CRISPRs)/CRISPR associated protein 9 (Cas9) is a prokaryotic immune system and has been used as an influential tool for genome editing (Ali et al. 2016). In the industrial biotechnology, this new technology has been introduced and editing of microbes like algae, fungi bacteria, and yeast is now possible through CRISPR/Cas9 technology (Ali et al. 2016; Borodina and Nielsen 2014). This chapter not only summarizes the conventional genome editing approaches but also encompasses the recent developments of CRISPR/Cas9-mediated genome editing in microorganisms for sustainable and high ethanol production.

5.2 Sources of LCB

Biomass or LCB is the world's largest renewable resource, which can be efficiently obtained from various organic materials. It has many useful components for ethanol production (Limayem and Ricke 2012). Forestland resources primarily consist of woody biomass, i.e., hardwoods and softwoods along with sawdust, pruning, and bark thinning remains, while grasslands contain agricultural remains. Industrial and municipal wastes are also considered as potential renewable cellulosic raw materials that can be produced either from domestic or non-domestic materials (Jhunjhunwala et al. 2008; Cardona et al. 2010). The LCB is commonly divided into five groups, i.e., forest woody feedstocks, municipal solid waste, agricultural residues, industrial waste, and microalgae (Saxena et al. 2015).

5.2.1 Forest Woody Feedstocks

Wood residues have historically remained a chief source of energy for 2000 million people around the world (Organization 2015). Forestry remains are comprised of residues including biomass, which has not been harvested by forest management processes. Forestry waste consists of sawdust, bark, and wood chips. Almost 65% of the biomass energy potential is obtained from it (Werther et al. 2000; Bååth et al. 2002). The distinguishing features of forest biomass such as low ash contents make it a very important choice over agricultural biomass for bioethanol production (Zhu and Pan 2010).

5.2.2 Municipal Solid Waste

A large quantity of Municipal Solid Waste (MSW) is collected and disposed of in open fields. The treatment and recycling of this MSW are very important in terms of environmental perspectives. Energy can be extracted from MSW through various processes. Many environmentally-friendly and economically viable technologies have been developed to produce ethanol from MSW (Hadar 2013). The MSW in various countries is divided into many portions, though the biodegradable organic contents of MSW are comprised of kitchen waste, paper, cardboard, and garden waste. However, these are not good raw materials due to the variety in MSW constituents and less than ideal source of separation and presence of contaminants (Mena-Violante and Olalde-Portugal 2007). However, it may serve as feedstock in those regions, which are deficient in appropriate raw materials. However, with the innovation in cellulosic ethanol generation, the Mediterranean's can utilize the cellulosic constituents of MSW as a transport fuel feedstock. It was concluded by Faraco and Hadar (2011) that if 60 to 90% realistic outcome for ethanol production is supposed, the Mediterranean can generate between 17 and 25 billion liters per year of ethanol from 50% of the 180 million tons of waste generated annually (Faraco and Hadar 2011). The lignocellulose is the organic portion of MSW, which can be a suitable raw material for ethanol production (Mena-Violante and Olalde-Portugal 2007).

5.2.3 Agricultural Residues

Agricultural residues are one of the bountiful and high in lignocellulose biomass source that is produced during or after the processing of crops that are available in large quantities for production of bioethanol (Saxena et al. 2015). In countries like Pakistan, the cultivation of various agricultural crops generates large quantities of waste that remains in the field or burnt (Bhutto et al. 2011). Thus, these residues can be utilized as a better substitute resource to produce biofuels such as bioethanol, in an economically approachable way. The utilization of agricultural residues for ethanol production reduces deforestation by minimizing the dependence on woody biomass. However, the harvest period of the crop residues is short (Limayem and Ricke 2012). The agricultural residues, as well as waste utilization for bioethanol generation, is a profitable and environment-friendly method to favor the development and generation of bioethanol from lignocellulosic agricultural residues (Bhatia and Paliwal 2011).

5.2.4 Industrial Waste

The organic components of industrial waste are an economic source of biomass for ethanol production. Industrial waste is a fourth-generation biofuel and the feedstocks of industrial waste are CO_2, capturing and recycling the carbon, water, etc. (Ahmed and Ahmaruzzaman 2016). The general potential of the organic contents of industrial waste is dependent on economic growth and utilization of biomaterials (Kang et al. 2014). Among several industries, palm oil, natural rubber, and biodiesel processing wastes are of certain interest due to their high potential organic loading and abundance. Potentially recyclable Cellulosic materials are also considered as industrial waste that originates either from inhabited or uninhabited sources like paper mill sludge and food wastes (Limayem and Ricke 2012). Biofuel that is gained from industrial waste will be more advanced as a sustainable source of energy in the near future. More study is required for the modernization of fourth-generation biofuel production on a commercial scale.

5.2.5 Microalgae

The use of microalgae to produce bioethanol is an economic and effective approach for bioethanol production using the LCB (Limayem and Ricke 2012). Increasing the efficacy of algae for ethanol production is preferred due to many advantages. Currently, marine algae biomass is an attractive feedstock for third-generation biofuel production due to the quick biorefineries development. Microalgae is not only a potential biomass for bioethanol, but it can also be a feedstock for other biofuels, specifically, biodiesel and fuel for aviation (Chernova and Kiseleva 2014). In contrast to corn and sugarcane, algal biomass can easily be cultivated as it does not require farmland and freshwater for its growth and it does not compete with food crops. Moreover, it utilizes high levels of CO_2 during growth, which makes it eco-friendly (Harel 2009).

5.3 Composition of Lignocellulosic Biomass

The LCB consists of three main components, namely, lignin, hemicellulose, and cellulose in addition to a minute quantity of other constituents, such as minerals, phenolic substituents, and acetyl groups. In lignocellulosic biomass, these polymers arranged themselves in complex three-dimensional structures. Structural complexity varies with the variation in the source of lignocellulosic feedstock (Isikgor and Becer 2015). Usually, lignocellulosic biomass contains 35–50% cellulose, 10–25% lignin, 20–35% hemicellulose, and fractions of oils, ash, and proteins (Saha et al. 2005). Cellulose is the main component of LCB and contains glucan polymers. It also has cellobiose (a disaccharide) as a repeating structural unit. Within its structure, the glucose units are closely bounded by intramolecular and intermolecular hydrogen bonding networks. Subsequently, cellulose constitutes about half of the total organic carbon present in the environment; the use of cellulose for making fuels and other valuable chemicals has paramount importance (Hendriks and Zeeman 2009; Stefanidis et al. 2014). The second most abundant polymer is the hemicelluloses, which has amorphous and random structure comprised of various heteropolymers including xyloglucan, arabinoxylan, glucomannan, galactomannan, xylan, and glucuronoxylan (Meng et al. 2015). The composition of hemicellulose is also different as hardwood hemicelluloses mostly consist of xylans, whereas softwood hemicelluloses generally comprise glucomannans (Stefanidis et al. 2014). The hemicelluloses are the heteropolymers that are composed of pentoses (like xylose and arabinose), hexoses (like mannose, glucose, and galactose), and acetylated sugars. A complex network of a bond is formed in the plant cell wall due to hemicelluloses increasing the strength by cross-linking with lignin. Conclusively, the three-dimensional structure of lignin is comprised of repeating units of phenylpropanoid. It is a cellular glue that provides stiffness to plant tissue and the individual fibers (Jönsson and Martín 2016; Isikgor and Becer 2015).

5.4 Bioconversion of LCB into Bioethanol

The biofuel production is directly linked with the yield of fermentable monomeric sugars obtained from LCB. The basic steps involved in the bioconversion of LCB into bioethanol are pretreatment, hydrolysis, fermentation, and product separation/distillation (Fig. 5.2).

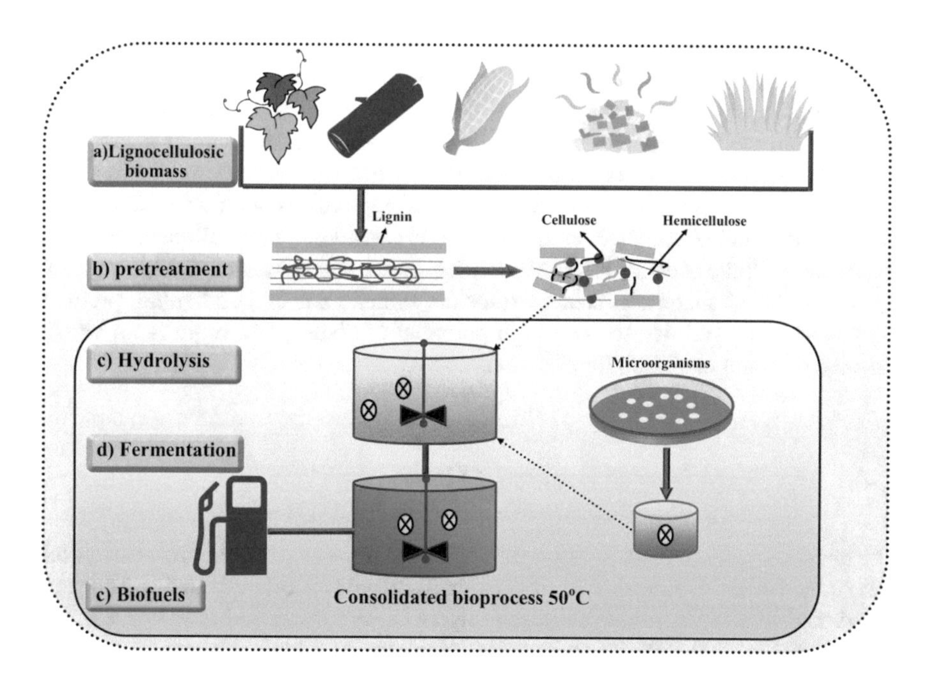

Fig. 5.2 Schematic representation of bioconversion of the LCB to ethanol

5.4.1 Pretreatment of LCB

The effective pretreatment of LCB is the necessary pre-requisite for optimal downstream processing of biomass into bioethanol (Limayem and Ricke 2012). Prior to pretreatment the raw material is disrupted physically and thermochemically resulting in increased porosity of the biomass, thus exposing the cellulose surface for the action of cellulase. This strategy helped to pave the way toward an effective saccharification approach with reduced energy input and increased sugar recovery (Zhu et al. 2009b; Zheng et al. 2009).

Various pretreatment approaches such as physical, chemical, and biological have been employed to disintegrate the tough chemical materials present within feedstock (Banerjee et al. 2010). Many efficient pretreatment methods have been developed during the last few years (Limayem and Ricke 2012). However, most of the pretreatment methods generate cellulose and hemicellulose from the LCB, but the key technical issues associated with some of the treatments render them economically inefficient. Moreover, all of the pretreatment methods are not capable to reduce the crystallinity of the material present in the woody feedstocks. In fact, some methods have the potential to overcome the feedstock variability (Klinke et al. 2004). The forest and woody raw materials, containing high lignin content (29%) and thus are more crystalline than agricultural feedstocks (like corn stover, rice, and wheat straws) having low lignin content (3–13%) Demirbas (2005).

The most commonly and widely used pretreatments include hot water, acid pretreatment, dilute acid hydrolysis, and lime. The implications of both dilute and concentrated acid pretreatments are limited by the production of toxic inhibiting compounds (Limayem and Ricke 2012). The high concentration of acetic acid production during dilute acid pretreatment is more difficult to separate and neutralize than furfural and HMF (Larsson et al. 1999). Another method named as ammonia fiber explosion (AFEX) has been developed for the efficient treatment of agricultural raw material (i.e., corn stover). The AFEX approach is not associated with the production of inhibitors such as furfural and is considered more efficient than acid treatments (Klinke et al. 2004). However, for the decomposition of highly tough and crystalline forest materials, steam explosion, organic solvent, and sulfite pretreatments are employed due to their high potential to disintegrate high lignin woody biomass (Öhgren et al. 2007b; Zhu et al. 2009a).

5.4.2 Hydrolysis of Pretreated LCB

The hydrolysis of pretreated LCB can be achieved in two ways either by acid hydrolysis or by enzymatic hydrolysis. Acid hydrolysis includes diluting acid and concentrated acid hydrolysis. Dilute acid hydrolysis (1–3%) is carried out at an extreme temperature of 200–240 °C to degrade cellulose. The high-temperature requirement and production of toxic inhibitors like HMF and phenolic compounds through the breakdown of pentoses and hexoses are the major limitations that impede its applications on a commercial scale.

To overcome the above-mentioned limitations, concentrated acid hydrolysis can be utilized to degrade the cellulose and hemicellulose into monomeric sugars. This method gives high sugar recovery (up to 90%) in contrast to the dilute acid method. However, the limitations that hinder its commercial-scale applications include large amounts of acid requirement, expensive acid recycling process, and equipment corrosion.

These issues have triggered the research efforts to develop efficient hydrolytic enzymes and enzymatic hydrolysis. For optimal enzymatic hydrolysis, an effective pretreatment method is required. In enzymatic hydrolysis, the depolymerization of LCB is performed by applying the active enzyme system. The enzyme-mediated LCB hydrolysis can be done either by using microbes that secretes LCB-degrading enzymes or by using a commercially available enzyme system. The latter one is more effective and widely used. Contrary to acid hydrolysis, low energy input and moderate conditions are required for enzymatic hydrolysis. The cellulase works efficiently at a temperature ranging from 40 to 50 °C and pH 4 to 5 (Neves et al. 2007). Park et al. (2002) reported that assay conditions for xylanase to be 50 °C temperature and pH 4–5. This is why, enzymatic hydrolysis due to its low toxicity, cost-effectiveness, and less corrosion than acid hydrolysis is an advantageous hydrolysis approach. Additionally, no inhibitory by-product has resulted during enzymatic hydrolysis (Sarkar et al. 2012).

5.4.3 Fermentation

The fermentation is the microbe-mediated transformation of the monomeric sugars into ethanol (Ocreto 2013). The upstream process of fermentation produces monomeric sugars, which are subsequently treated by numerous microbes for ethanol production. But the lack of ideal strain to ferment pentoses and hexoses effectively limits the implication of LCB for bioethanol production on a large scale (Talebnia et al. 2010). The most effective ethanol-producing technique on a commercial scale is one that involves an ideal microorganism with broad substrate range, can produce ethanol in high concentrations, can tolerate high ethanol concentration and heat, can be able to grow in the presence of chemical inhibitors produced during hydrolysis process and have cellulolytic activity. To achieve these goals genetic modifications of microorganisms have gained attention to develop strains that have the ability to utilize sugars completely released during pretreatment and hydrolysis and to provide better ethanol yield (Sarkar et al. 2012). The extensively used microbial species in the fermentation process include *Zymomonas mobilis*, *Pachysolen tannophilus*, *Pichia stipitis*, *Escherichia coli*, *Saccharomyces cerevisiae*, *Mucor indicus*, *Candida shehatae*, and *Candida brassicae* (Sarkar et al. 2012).

Various fermentation techniques are implemented to convert lingo-cellulosic hydrolysate into bioethanol such as Separate Hydrolysis and Fermentation (SHF), and Simultaneous Saccharification and Fermentation (SSF). The problems related to this approach are retardation of enzyme activity by inhibitory end-product and contamination. The drawbacks of the SHF process can be handled with the development of the SSF approach in which a single container is utilized for simultaneous hydrolysis and fermentation. Carbohydrates formed after hydrolysis are fermented into ethanol immediately, and hence, are involved in running the process smoothly by avoiding carbohydrate accumulation (Öhgren et al. 2007a). Moreover, in contrast to SHF, SSF is less expensive as it requires a single reactor for both hydrolysis and fermentation. The major constraint limiting the application of SSF on a commercial scale is the difference in the optimum temperature of hydrolysis and fermentation. Several yeasts involved in fermentation process work effectively on optimal temperature ranging from 30 to 35 °C, whereas hydrolytic enzymes function efficiently on 50 °C temperature (Kádár et al. 2004). The SSF limitation can be overcome by using thermo-tolerant microbes, e.g., *Kluyveromyces marxianus* and *Saccharomyces cerevisiae* MLD10 which have been developed to tolerate high temperatures required for enzyme-mediated hydrolysis (Bjerre et al. 1996; Arshad et al. 2014).

The available alternatives of SSF or SHF are Direct Microbial Conversion (DMC) or Consolidated Bioprocessing (CBP) and Simultaneous Saccharification and Co-Fermentation (SSCF) Cardona et al. (2010). In CBP approach, only a single reactor is utilized for carrying out all processes like cellulase production, biomass hydrolysis, and ethanol fermentation that are involved in the biotransformation of LCB in an integrated manner (Zhao et al. 2011; Kang et al. 2014). Cellulose can be fermented into ethanol directly by applying mono- or mixed-culture of microorganisms. The CBP is a cost-effective approach as it does not require capital investment for enzyme purchasing and production (Lynd et al. 2002). A bacterial species named *Clostridium thermocellum* and various fungal species such as *Neurospora crassa, Fusarium oxysporum,* and *Paecilomyces* sp. have been revealed to have this type of activity. However, CBP process results in poor yields of ethanol and time taking fermentation periods (3–12 days), which make this process inefficient (Szczodrak and Fiedurek 1996). The co-fermenting microbial cultures in SSCF are required to be compatible in terms of operating temperature and pH. Neves et al. (2007) revealed that the consortium of *Saccharomyces cerevisiae* and *Candida shehatae* is suitable for the SSCF approach. In order to ferment sugars with maximum efficacy, the fermentation process has been done in an ordered manner involving *S. cerevisiae* application for hexose utilization followed by *C. shehatae* inoculation for the conversion of pentoses, but the poor yield of ethanol is obtained by using this sequential approach of fermentation (Sanchez and Cardona 2008).

5.4.4 Product Purification

Bioethanol obtained after the fermentation process is further purified by using the distillation process. The ethanol is separated from the water on the basis of its volatility by employing fractional distillation process (Cardona and Sánchez 2007).

In the first step, the distillation of fermentation products is carried out in order to separate ethanol from water and the remaining liquid is transmitted to the wastewater treatment (Zheng et al. 2009). All components in left-over liquid may be concentrated and can be either used to provide power to run the process as burning fuel or transformed into other co-products (Zabed et al. 2016).

5.5 Microbial Role in Ethanol Production from LCB

The variety of microorganisms belonging to diverse groups, e.g., fungi, yeast, actinomycetes, and bacteria have a pivotal role in the biotransformation of LCB into bioethanol. The potential roles and characteristics of various microbes are summarized in Table 5.1.

5.5.1 Role in Pretreatment and Detoxification

Biological pretreatment of LCB releases the components from the lignocellulosic complex for further processing. The soft rot and white rot fungi having the potential enzymatic system are extensively used for this purpose. These fungi disintegrate feedstock by producing lignin peroxidase (Zabed et al. 2016).

Similarly, the hydrolysate detoxification process includes microbes producing peroxidase and laccase enzymes that help in the removal and neutralization of inhibitors (Chandel et al. 2013). The effective bacterial strains for the detoxification process include *Pseudomonas* sp., *Actinobacter* sp., *Methylobacterium extorquens*, *Arthrobacter aurescens,* and *Flavobacterium indologenes*. The fungal strain *Coniochaeta ligniaria* C8 NRRL30616 has been found to be very efficient in the neutralization of inhibitors such as ferulic acid and furfural (Chandel et al. 2013; López et al. 2004). The derivatives of benzoic acid, acetic acid, and furfural present in the hydrolysate have been removed by treating with *Trichoderma reesei* (Palmqvist and Hahn-Hägerdal 2000).

5.5.2 Role in Hydrolysis

The enzymatic hydrolysis has become the method of choice due to reduced energy and requirement of moderate processing conditions in contrast to chemical hydrolysis (Ferreira et al. 2009). The enzymatic hydrolysis of cellulose includes the microbial species producing hydrolytic enzymes like cellulases (Sun and Cheng 2002). The bacterial genera reported for degradative enzyme production include *Clostridium, Bacteroides, Erwinia, Bacillus, Cellulomonas, Ruminococcus, Acetovibrio, Streptomyces, Microbispora,* and *Thermomonospora*. Whereas potential fungal genera

Table 5.1 Characteristics and processes of the microorganisms involved in lignocellulosic-based bioethanol production

Microbe	Biochemical process	Key role/characteristics	Reference
Bacteria			
Xanthomonas sp., *Pleurotus ostreatus*, *Pseudomonas* sp., *Phanerochaete chrysosporium*, *Thermomonospora mesophila*, *Acinetobacter* sp.	Pretreatment	• These bacterial species have the potential to degrade lignin and cellulose via producing ligninases and cellulases	Rabemanolontsoa and Saka (2016), Singh et al. (2008)
Caldibacillus cellulovorans, *Caldocellum saccharolyticum*	Hydrolysis	• Able to produce mannanase	Gírio et al. (2010), Sunna et al. (2000)
Bacillus subtilis, *Cellulomonas biazotea*	Hydrolysis	• Involved in cellulolytic activity	Saratale et al. (2010), Menon and Rao (2012)
Thermotoga maritima, *Clostridium cellulolyticum*, *Clostridium cellulovorans*, *Sulfolobus solfataricus*, *Alicyclobacillus acidocaldarius*	Hydrolysis	• Display cellulolytic and hemicellulolytic activities	Rabemanolontsoa and Saka (2016)
Zymomonas Mobilis	Fermentation	• Higher ethanol yield than *Saccharomyces cerevisiae* (five times greater compared to yeast) • Can survive under high ethanol production (i.e., up to 14% v/v)	Gírio et al. (2010), Balat et al. (2008)
Escherichia coli	Fermentation	• Have the potential to ferment both pentoses and hexoses • Easy to modify genetically	Qureshi et al. (2012), Dien et al. (2000)
Klebsiella oxytoca	Fermentation	• Can ferment various sugars efficiently such as pentoses, hexoses, cellobiose, and cellotriose • High fermentation rate than yeasts	Balat (2011), Hayes (2009)
Pachysolen tannophilus	Fermentation	• Can ferment pentose and hexose sugars simultaneously	Baeyens et al. (2015)

(continued)

Table 5.1 (continued)

Microbe	Biochemical process	Key role/characteristics	Reference
Yeasts			
Clavispora NRRL Y-50464	Hydrolysis and fermentation	• Display good tolerance to inhibitors • Cellobiase producer	Liu et al. (2012)
Saccharomyces cerevisiae	Fermentation	• High ethanol production (i.e., 90%) • Can ferment hexoses like glucose, galactose, and mannose as well as disaccharides such as sucrose and maltose Have good resistance against chemical inhibitors and ethanol concentration (i.e., 10% v/v) • Easy genetic manipulations	Nigam (2002), Gírio et al. (2010), Almeida et al. (2007), Arshad et al. (2008)
Brettanomyces naardenensis, Pachysolen tannophilus, Pichia stipites, Pichia segobiensis, Candida tenuis, Candida shehatae	Fermentation	• Well suited for xylose fermentation • Capable of neutralizing the detrimental effects of chemical inhibitors such as furfural and hydroxymethylfurfural • Can also ferment cellobiose by producing cellobiase	Silva et al. (2014), Almeida et al. (2008), Silva et al. (2016), Gírio et al. (2010), Balat (2011), Slininger et al. (2015)
Kluveromyces marxianus	Fermentation	• Capable to ferment wide variety of sugars	Baeyens et al. (2015), Wu et al. (2016)
Fungi			
Aspergillus terreus	Pretreatment	• Able to degrade lignin and cellulose	Singh et al. (2008)
Pleurotus florida, Ceriporiopsis subvermispora, Phanerochaete chrysosporium	Pretreatment	• Involved in delignification of the LCB	Rouches et al. (2016)

(continued)

Table 5.1 (continued)

Microbe	Biochemical process	Key role/characteristics	Reference
Thermoascus aurantiacus	Hydrolysis	• Reported to produce xylanase as well as cellulase	Zabed et al. (2016), Kalogeris et al. (2003)
Fusarium oxysporum, Trichoderma viridae, Penicillium funiculosum, Aspergillus terreus M11	Hydrolysis	• Cellulase production	Kalogeris et al. (2003), Adsul et al. (2004), Pham and Halos (1990), Panagiotou et al. (2003
Humicola insolens, Termomonospora fusca, Aspergillus niger, Trichoderma longibrachiatum, Trichoderma reesei	Hydrolysis	• Able to produce xylanase	Gírio et al. (2010)

include *Aspergillus*, *Penicillium*, *Phanerochaete*, *Schizophyllum*, *Sclerotium*, *Trichoderma*, *Humicola*, *Schizophillum,* and *Fusarium* (Balat 2011; Gupta and Verma 2015). The hemicellulose due to its complex and tough structure requires an array of hydrolytic enzymes for its degradation in contrast to cellulose. Numerous fungal and bacterial genera such as *Penicillium*, *Talaromyces*, *Trichoderma*, *Aspergillus*, and *Bacillus* have been reported to produce a set of hemicellulose degradative enzymes (Balat 2011).

5.5.3 Role in Fermentation

Numerous microbes belonging to different groups, like yeast, filamentous fungi, and bacteria, convert fermentable carbohydrates into ethanol under hypoxic conditions. The filamentous fungi cannot produce ethanol in higher concentrations because of its reduced tolerance to ethanol. It has been reported that every microbe has different ethanol-producing potential. The industrial-scale ethanol production involves only a few species which are capable of producing ethanol in high concentrations. Different factors to be considered for enhanced ethanol production are presented in Fig. 5.3.

Most of the industrial processes commonly used *Saccharomyces cerevisiae* (a yeast) for ethanol production due to its robust and suited nature for the fermentation of soluble sugars. The bacterial species include gram-negative bacterium, e.g., *Zymomonas mobilis* and various anaerobic species like *Clostridium thermosaccharolyticum*, *Thermoanaerobacter mathranii*, *Clostridium thermohydrosulfuricum*, *Thermoanaerobium brockii*, *Thermoanaerobacter ethanolicus,* etc.

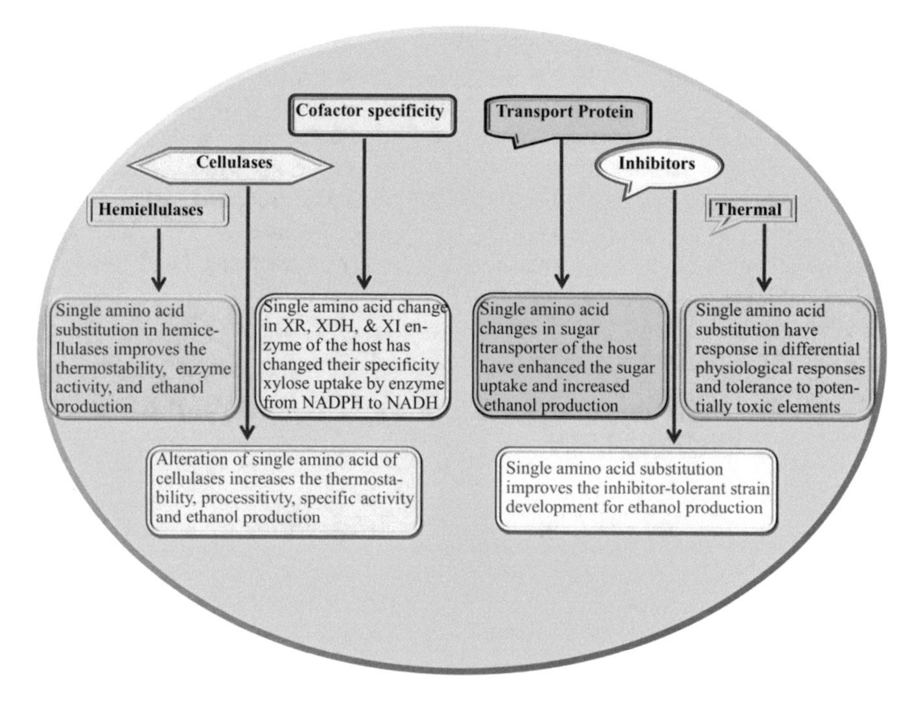

Fig. 5.3 Different factors to be considered for enhanced ethanol production

5.6 Genome Engineering: The Dawn of the Modern Genomics Era

Genome engineering is an efficient way to alter the genome of an organism to obtain beneficial and desirable characters. It alters the genome of an organism in a targeted manner with high precision to produce the organism with desirable physiological traits (Singhania et al. 2017; Ulaganathan et al. 2017). This tool can also be utilized for microbial genome modifications to promote the production of certain metabolites (Singhania et al. 2017). The specific gene in an organism can be inserted, deleted, and up or down-regulated by using this process. This approach outclassed the conventional genome modification techniques in many aspects such as no need to isolate a gene of interest, in vitro engineering and re-introduction of the recombinant molecule into the organism to induce the desirable physiological traits (Ulaganathan et al. 2017). There are two approaches utilized for genome modification: one is RNA guided Endonuclease Mediated (REM) engineering and second is Modified Endonuclease Mediated (MEM) engineering. The clustered regularly interspaced short palindromic repeats (CRISPR)/CRISPR associated protein 9 (Cas9) system mediated genome engineering is REM type (Wiedenheft et al. 2012), whereas MEM technique involves transcription activator-like effector nucleases (TALENs) system

(Miller et al. 2011), zinc finger nucleases (ZFNs) system (Kim et al. 2011), and meganucleases system (Smith et al. 2006) of genome engineering.

The dawn of such genome engineering approaches has revolutionized the research studies in biology and other fields of science. Today one of the important questions is which approach is most optimal in genome engineering. Various factors, viz., unavailability of suitable and efficient delivery, less precision, and toxicity are the major limiting factors for ZFNs and TALENs. The emergence of the CRISPR/Cas9 system of genome engineering has reduced the limitations associated with ZFNs and TALENs (Gaj et al. 2013).

5.7 Genome Engineering for Enhanced Ethanol Production Using CRISPR-Cas9

5.7.1 Engineering for Lignocellulose Utilization

The cellulose and hemicellulose of LCB can be hydrolyzed by microbe-oriented cellulolytic and hemicellulases into monomeric sugars, respectively. Cellulolytic enzymes consist of different types of cellulases like endoglucanases, exoglucanase or cellobiohydrolase, β-glucosidase, acetylesterase, glucuronidase, β-xylosidase, galactomannanase, xylanase, and glucomannanase that work on cellulose synergistically and convert it into glucose. Out of these only three types of cellulase enzymes, viz., carboxymethylcellulases or endo-1-4-β-glucanase (EC 3.2.1.4), cellobiohydrolase (EC 3.2.1.91), and β-glucosidase (EC 3.2.1.21) are required for the complete hydrolysis of cellulose into glucose (Lynd et al. 2002). The endoglucanase attacks low crystallinity regions within the cellulose structure producing oligosaccharides, and the oligosaccharides are further hydrolyzed by cellobiohydrolase into cellobiose units. The resulting units are attacked by β-glucosidase and converted into glucose (Lynd et al. 2002; Bhatia et al. 2017). Most microbes lack the potential to disintegrate and utilize lignocellulosic material, and their potential can be improved through metabolic engineering (Wagaba et al. 2016). Extensive research efforts are in pipeline to modify microbial genome to produce engineered cellulosomes for efficient degradation of polymeric sugars. Morais and coworkers found that a combination of thermophilic enzymes and scaffoldin is more effective for cellobiose digestion. An accelerated cellobiose utilization and fermentation were observed in *Saccharomyces cerevisiae* overexpressing the cellobiose transporter and β-glucosidase gene, resulting in 0.42 g ethanol production per gram cellobiose utilization (Wagaba et al. 2016; Kim et al. 2013).

5.7.2 Metabolic Engineering of Microorganisms for Biofuel Production

Biofuel production can be enhanced by metabolically engineered microorganisms that may have more productivity, product diversity, yield, and concentration. Several microorganisms like bacteria, yeast, cyanobacteria, and microalgae have the capability to convert numerous substrates into ethanol. Molecular biology of two bacterial species, i.e., *Bacillus subtilis* and *Escherichia coli* is very well-understood, so, these species are most widely used to construct strains that can produce ethanol. These bacteria can utilize various substrates and are acquiescent for metabolic engineering to produce ethanol exclusively (Bhatia et al. 2016). Through metabolic engineering, the first microorganism that had been successfully improved to produce ethanol was probably *E. coli* (Zhou et al. 2005). Biofuels, particularly bioethanol can be produced by using yeasts that are known as cell factories. The yeast *S. cerevisiae* is the best-known yeast that also has a long history of industrial use has attained special interest (Dalawai et al. 2017). Ethanol titer of 1.4 g/L can be achieved by simultaneous saccharification and fermentation of crystalline cellulose because *S. cerevisiae* was engineered to surface display mini-cellulosomes. Additionally, *S. cerevisiae* was engineered for the fermentation of that cannot be fermented previously by it. It was investigated that an engineered strain, which has the capacity to co-ferment the mixture of cellobiose and xylose and it can produce ethanol nearly 0.65 g/L/h (Hayat et al. 2012; Cheng et al. 2011). Some of the cyanobacteria have the natural ability to produce pure hydrocarbons and alkanes, energy-rich potential fuels are alkanes. The new option can be explored in the future for the production of fuels by using engineered cyanobacteria (Schirmer et al. 2010). The microalgae have gained importance as manufacturers of ethanol. Future commercialization of algal fuels can be achieved by metabolic and genetic engineering.

5.7.3 Engineering for Thermo-Tolerance Development

Utilization of the heat-stable enzymes/organisms in biological conversion of the lignocellulosic biomass allows better suppleness in the complete bioprocess because of (i) drop in the cost of production, (ii) augmented stability and definite action of the enzymes, (iii) swelling of cellulose allows easy breakdown of it at raised temperature, (iv) inhibition of growth of contamination by microorganisms, (v) fluid viscosity is reduced that leads to improved mass transfer proportion, and (vi) non-enzymatic pretreatment methodologies are applied for increasing compatibility (Ulaganathan et al. 2017). The manufacturing of bioethanol usually at a high temperature can lessen the viability of microorganisms which produce ethanol. There is a wide range of substances and proteins present that can help in achieving thermotolerance, for example, a protein capable of bearing heat shock and secretion of compound trehalose, which can confer thermostability. In *S. cerevisiae* and *Z.*

mobilis, ethanol production is improved and thermotolerance is achieved by changing a single amino acid in C-5 sterol desaturase, NADH dehydrogenase, and pyruvate kinase (Caspeta et al. 2014). Enhanced heat tolerance is gained by deleting the Dfg5 glycosylphosphatidylinositol an anchored member protein gene in *S. cerevisiae*. The cellulose hydrolysis effectiveness can be dramatically increased by using *T. aurantiacus* that produce thermostable xylanase when it is added before the hydrolysis. Not only the removal of the hemicellulose but also xylanase binding on both lignin and hemicellulose can lessen the unnecessary binding of cellulase can enhance the hydrolysis (Monte et al. 2010). A wide range of amino acid replacement has been found by working on cellulase and hemicellulose that resulted in the thermostability of the enzymes and half-life of the enzymes is increased at high temperatures and this procedure involves techniques of either protein engineering or evolutionary engineering.

5.7.4 Engineering for Cofactor Specificity

Cofactor specificities of enzymes can be altered by applying genome engineering in the xylose utilization pathway. In organisms such as *Gluconobacter oxydans*, *Pichia stipitis*, *Piromyces* sp., *S. cerevisiae*, *Aspergillus carbonarius*, and *Candida tenuis* the cofactor specificities of enzymes can be changed by amino acid change in enzymes like xylitol dehydrogenase, xylose isomerase, and xylose reductase (Ulaganathan et al. 2017). The xylose uptake can be improved in *S. cerevisiae* as a result of the deletion of the FPS1 gene (Kim et al. 2013). The pathway engineering of *S. cerevisiae* has been successfully achieved by CRISPR/Cas system by precisely inserting multiple genes in the selected genomic location (Horwitz et al. 2015) that is responsible for the precise engineering in *S. cerevisiae* for modifying xylose utilization pathways. In bifunctional ADH/ALDH enzyme, single amino acid replacement (G544D) results in enhanced NADPH specificity in *Thermoanaerobacter saccharolyticum* (Ulaganathan et al. 2017).

5.7.5 Engineering for Enhancing Ethanol Tolerance

Biofuels can be produced by utilizing a biosynthetic pathway present in microbes. Yield and productivity are affected by the biofuel product itself because it possesses antimicrobial activity. So, such microbes are needed to produce by engineering techniques that have improved tolerance for end-product (Dunlop 2011). In different strains like *E. coli*, *Z. mobilis*, and *S. cerevisiae* variation in the tolerance of biofuel have been observed. However, ethanol tolerance is a multifaceted attribute connecting numerous networks or pathways but transcript profiling analyses and comparative functional genomics in *S. cerevisiae* have recognized the genes that are responsible for the tolerance of ethanol (Lewis et al. 2010; Ma and Liu 2010a, b). In *Clostridium thermocellum*, ethanol tolerance is raised up by point mutation of the proline residue at 704 positions that is replaced with leucine and histidine residue at 734 positions which is changed with arginine in the alcohol dehydrogenase enzyme. This increase is because of the alteration in the requirement of the cofactor of enzyme alcohol dehydrogenase (Brown et al. 2011). Many of the approaches have been used for decreasing the intrinsic toxicity of biofuel, to counter sensitivity. These approaches include stimulation of stress-responsive genes, the heterologous expression of heat shock proteins or expression of efflux pumps, and alteration of membrane proteins (Dunlop 2011). The heat shock proteins have an important role in manufacturing, confirming the precise folding of proteins and transport. During the stress conditions, these proteins respond by up-regulating and play an essential part in averting protein accumulation. Rutherford and colleagues did experiments of expression profiling on *E. coli* cells that are exposed to butanol solution and they identified many of the affected genes (*htpG, ibpAB, rpoH,* and *dnaJ*) which are responsible for misfolding of proteins and heat shock (Rutherford et al. 2010). The microbes alter the composition of the cell wall during solvent exposure to them, in this way they prevent the solvent entry inside the cell. By using enzyme cis-trans isomerase, microbes can modify *cis* fatty acids and form *trans* fatty acids. Modification in *trans* fatty acids to form *cis* fatty acids proportion may have the ability to transform the membrane fluidity and it may affect solvent tolerance and permeability (Bhatia et al. 2017).

5.8 Concluding Remarks and Future Prospects

The fuel demand and consumption are expected to increase rapidly and fossil energy utilization will pose serious harm to the environment (Arshad 2017). According to the International Energy Agency (IEA), an integrated approach is required to provide the world with better energy security and energy access. Such goals can be achieved by aligning the energy efficiency and renewable energy policies. The current global energy crisis can only be resolved with the development of efficient and sustainable energy supplies. It has been reported that one farm in Tunisia has developed a form of biofuel as an alternative to firewood, which many consider as major environmental hazard, to meet up the challenges due to increased energy scarcity (Chanouf Farm Biofire 2018). Thus, worldwide energy issues can be resolved by switching toward biofuels as an alternative to fossil energy. Biofuels or biomass-derived fuels can be produced by the action of potential microorganisms (Voloshin et al. 2016). Today, bioethanol is one of the most abundantly consumed biofuels (Voloshin et al. 2016). The eco-friendly nature and energy sustainability made biofuels advantageous over conventional fuels (Saini et al. 2015).

During the past few years, plenty of investigations has been done for designing and development of the bioconversion process to convert LCBs into biofuel in an efficient manner. For commercial-scale production of biofuel, various strategies or configurations such as SHF, SSF, SSCF, and CBP have been designed (Saini et al. 2015). The key steps of these processes involve microbial species having the tendency to derive the process efficiently (Zabed et al. 2016). Each technique is limited by certain factors, which will ultimately affect biofuel production. These limitations include end-product enzyme inhibition, microbial sensitivity toward biofuel concentrations, temperature, and pH (Saini et al. 2015). Genetically modified microbes have drawn the attention of the scientific community as the best alternative to these limitations to bio-transform low-cost LCBs into biofuel in an efficient manner (d'Espaux et al. 2015). Numerous reports are available describing the role of genetically modified microbes in enhanced biofuel production (d'Espaux et al. 2015; Gonçalves and Simões 2017; Ulaganathan et al. 2017).

Numerous techniques of genome engineering like ZFNs, TALENs, and the CRISPR/Cas9 system (Shin et al. 2016) have been developed for the modification of genes. Among all techniques, CRISPR/Cas9 is the best-suited strategy for the genome modifications of microbial cells. Several reports have been published on the CRISPR/Cas9 system improving the characteristics of microbial cells (Kleinstiver et al. 2015; Gao et al. 2015). CRISPR/Cas9 may also serve as an efficient tool to modify biofuel producing microbial cells for enhanced biofuel production by targeting specific genes. Various bacterial species such as *Escherichia coli*, *Zymomonas mobilis*, and *Clostridium cellulovorans* with improved thermostability, inhibitors (like furfural) tolerance, biofuel tolerance, and production can be developed via CRISPR/Cas9 mediated genome manipulations by targeting specific genes. Various characters of *Saccharomyces cerevisiae* and fungal species like *Aspergillus usamii* and *Aspergillus niger* can be improved by genome engineering through the CRISPR/Cas9 approach for enhanced biofuel production.

However, one major constraint in biofuel production is the development of microbes with improved characteristics. The second limitation hindering biofuel production on a commercial level is the development of an efficient process configuration. So, plenty of research insight is necessary for the development and commercialization of biofuels. The use of LCB and microbial genome modification for enhanced bioethanol production will ultimately help to achieve the United Nations sustainable development goal "ensure access to affordable, reliable, sustainable, and modern energy for all".

Despite the enthusiastic views on the potential of biofuels for sustainable development, there is very little research on the links between biofuel production, trade, and sustainable development. Existing research focuses on the economic and technological aspects of biofuel production. Research on environmental aspects tends to concentrate on their energy balance and the potential for reduced greenhouse gase emissions. Almost no research has been done on the trade aspects or the wider implications for sustainable development of trade in biofuels. The major barrier toward the use of renewable energy source depends on a country's policy and policy instrument, which in turn affect the cost and technological innovations. In addition, technological innovations affect the cost of renewable energy technologies, which in turn leads to market failures, and low patronization of the renewable energy technology. In view of the above discussion, effective renewable energy policy should take the interconnection of factors affecting renewable energy supplies and sustainability into consideration.

References

Adsul M, Ghule J, Singh R, Shaikh H, Bastawde K, Gokhale D et al (2004) Polysaccharides from bagasse: applications in cellulase and xylanase production. Carbohydr Polym 57(1):67–72

Ahmed MJK, Ahmaruzzaman M (2016) A review on potential usage of industrial waste materials for binding heavy metal ions from aqueous solutions. J Water Process Eng 10:39–47

Ali Z, Ali S, Tashkandi M, Zaidi SS-e-A, Mahfouz MM (2016) CRISPR/Cas9-mediated immunity to geminiviruses: differential interference and evasion. 6: 26912. https://doi.org/10.1038/srep26912. https://www.nature.com/articles/srep26912#supplementary-information

Almeida JR, Modig T, Petersson A, Hähn-Hägerdal B, Lidén G, Gorwa-Grauslund MF (2007) Increased tolerance and conversion of inhibitors in lignocellulosic hydrolysates by *Saccharomyces cerevisiae*. J Chem Technol Biotechnol 82(4):340–349

Almeida JR, Modig T, Röder A, Lidén G, Gorwa-Grauslund MF (2008) *Pichia stipitis* xylose reductase helps detoxifying lignocellulosic hydrolysate by reducing 5-hydroxymethyl-furfural (HMF). Biotechnol Biofuels 1(1):12

Arshad M (2010) Bioethanol: A sustainable and environment friendly solution for Pakistan. Sci Vis 16(1 & 2):25–30

Arshad M (2017) Clean and sustainable energy technologies. In: Rasul M (eds) Clean energy for sustainable development, pp 73–89

Arshad M, Ahmed S, Zia MA, Rajoka MI (2014) Kinetics and thermodynamics of ethanol production by *Saccharomyces cerevisiae* MLD10 using molasses. Appl Biochem Biotechnol 1–10

Arshad M, Hussain T, Iqbal M, Abbas M (2017) Enhanced ethanol production at commercial scale from molasses using high gravity technology by mutant *S. cerevisiae*. Braz J Microbiol 48(3):403–409

Arshad M, Khan ZM, Khalil-ur-Rehman SFA, Rajoka MI (2008) Optimization of process variables for minimization of by product formation during fermentation of blackstrap molasses to ethanol at industrial scale. Lett Appl Microbiol 47:410–414

Arshad M, Zia MA, Shah FA, Ahmad M (2018) An overview of biofuel. In: Arshad M (eds. Perspectives on water usage for biofuels production. Springer, Cham

Bååth H, Gällerspång A, Hallsby G, Lundström A, Löfgren P, Nilsson M et al (2002) Remote sensing, field survey, and long-term forecasting: an efficient combination for local assessments of forest fuels. Biomass Bioenergy 22(3):145–157

Baeyens J, Kang Q, Appels L, Dewil R, Lv Y, Tan T (2015) Challenges and opportunities in improving the production of bio-ethanol. Prog Energy Combust Sci 47:60–88

Balat M (2011) Production of bioethanol from lignocellulosic materials via the biochemical pathway: a review. Energy Convers Manag 52(2):858–875

Balat M, Balat H, Öz C (2008) Progress in bioethanol processing. Prog Energy Combust Sci 34(5):551–573

Banerjee S, Mudliar S, Sen R, Giri B, Satpute D, Chakrabarti T et al (2010) Commercializing lignocellulosic bioethanol: Technology bottlenecks and possible remedies. Biofuel Bioprod Bior 4(1):77–93

Bhatia L, Paliwal S (2011) Ethanol producing potential of *Pachysolen tannophilus* from sugarcane bagasse. Int J Biotechnol Bioeng Res 2(2):271–276

Bhatia SK, Kim SH, Yoon JJ, Yang YH (2017) Current status and strategies for second generation biofuel production using microbial systems. Energy Convers Manag 148:1142–1156

Bhatia SK, Lee BR, Sathiyanarayanan G, Song HS, Kim J, Jeon JM et al (2016) Biomass-derived molecules modulate the behavior of *Streptomyces coelicolor* for antibiotic production. 3 Biotech 6(2):223

Bhutto AW, Bazmi AA, Zahedi G (2011) Greener energy: issues and challenges for Pakistan—biomass energy prospective. Renew Sustain Energy Rev 15(6):3207–3219

Bjerre AB, Olesen AB, Fernqvist T, Plöger A, Schmidt AS (1996) Pretreatment of wheat straw using combined wet oxidation and alkaline hydrolysis resulting in convertible cellulose and hemicellulose. Biotechnol Bioenergy 49(5):568–577
Borodina I, Nielsen J (2014) Advances in metabolic engineering of yeast *Saccharomyces cerevisiae* for production of chemicals. Biotechnol J 9(5):609–620
Brown SD, Guss AM, Karpinets TV, Parks JM, Smolin N, Yang S et al (2011) Mutant alcohol dehydrogenase leads to improved ethanol tolerance in *Clostridium thermocellum*. Proc Natl Acad Sci 108(33):13752–13757
Cardona C, Quintero J, Paz I (2010) Production of bioethanol from sugarcane bagasse: status and perspectives. Bioresour Technol 101(13):4754–4766
Cardona CA, Sánchez ÓJ (2007) Fuel ethanol production: process design trends and integration opportunities. Bioresour Technol 98(12):2415–2457
Caspeta L, Caro-Bermúdez MA, Ponce-Noyola T, Martinez A (2014) Enzymatic hydrolysis at high-solids loadings for the conversion of agave bagasse to fuel ethanol. Appl Energy 113:277–286
Chandel AK, Da Silva SS, Singh OV (2013) Detoxification of lignocellulose hydrolysates: biochemical and metabolic engineering toward white biotechnology. Bioenergy Res 6(1):388–401
Chang WR, Hwang JJ, Wu W (2017) Environmental impact and sustainability study on biofuels for transportation applications. Renew Sustain Energy Rev 67:277–288
Cheng NH, Liu JZ, Liu X, Wu Q, Thompson SM, Lin J et al (2011) *Arabidopsis* monothiol glutaredoxin, AtGRXS17, is critical for temperature-dependent postembryonic growth and development via modulating auxin response. J Biol Chem 286(23):20398–20406
Chanouf Farm Biofire (2018) Impact Journalism Day. https://www.un.org/sustainabledevelopment/blog/2018/07/chanouf-farm-biofire/
Chernova N, Kiseleva S (2014) Efficiency of the biodiesel production from microalgae. Therm Eng 61(6):399–405
Coyle W (2007) The future of biofuels: a global perspective. Amber Waves 5(5):24
d'Espaux L, Mendez-Perez D, Li R, Keasling JD (2015) Synthetic biology for microbial production of lipid-based biofuels. Curr Opin Chem Biol 29:58–65
Dalawai N, Krupa K, Nadkarni S, Bharani S, Harinikumar K (2017) Screening of efficient ethanol tolerant yeast strain for production of ethanol. Int J Pure App Biosci 5(1):744–752
Demirbas A (2005) Bioethanol from cellulosic materials: a renewable motor fuel from biomass. Energy Sourc 27(4):327–337
Dien BS, Nichols NN, O'bryan PJ, Bothast RJ (2000) Development of new ethanologenic *Escherichia coli* strains for fermentation of lignocellulosic biomass. Appl Biochem Biotechnol 84(1–9):181–196
Dunlop MJ (2011) Engineering microbes for tolerance to next-generation biofuels. Biotechnol Biofuels 4(1):32
Enhancing the Impact of Energy Efficiency and Renewable Energy Policies. International Energy Agency (2018). https://www.iea.org/newsroom/news/2018/march/enhancing-the-impact-of-energy-efficiency-and-renewable-energy-policies.html
Faraco V, Hadar Y (2011) The potential of lignocellulosic ethanol production in the mediterranean basin. Renew Sustain Energy Rev 15(1):252–266
Ferreira S, Duarte AP, Ribeiro MH, Queiroz JA, Domingues FC (2009) Response surface optimization of enzymatic hydrolysis of *Cistus ladanifer* and *Cytisus striatus* for bioethanol production. Biochem Eng J 45(3):192–200
Fulton L, Howes T, Hardy J (2004) Biofuels for transport: an international perspective. International Energy Agency, Paris, France
Gaj T, Gersbach CA, Barbas CF III (2013) ZFN, TALEN, and CRISPR/Cas-based methods for genome engineering. Trends Biotechnol 31(7):397–405
Gao J, Wang G, Ma S, Xie X, Wu X, Zhang X et al (2015) CRISPR/Cas9-mediated targeted mutagenesis in *Nicotiana tabacum*. Plant Mol Biol 87(1–2):99–110

Garneau JE, Dupuis ME, Villion M, Romero DA, Barrangou R, Boyaval P et al (2010) The CRISPR/Cas bacterial immune system cleaves bacteriophage and plasmid DNA. Nature 468(7320):67–71

Ghosh P, Westhoff P, Debnath D (2019) Biofuels, food security, and sustainability. In: Biofuels, bioenergy and food security. Academic Press, pp 211–229

Gírio FM, Fonseca C, Carvalheiro F, Duarte LC, Marques S, Bogel-Łukasik R (2010) Hemicelluloses for fuel ethanol: a review. Bioresour Technol 101(13):4775–4800

IPCC (2018) Report on global warming of 1.5 °C

Gonçalves AL, Simões M (2017) Metabolic engineering of *Escherichia coli* for higher alcohols production: an environmentally friendly alternative to fossil fuels. Renew Sustain Energy Rev 77:580–589

Gupta A, Verma JP (2015) Sustainable bio-ethanol production from agro-residues: a review. Renew Sustain Energy Rev 41:550–567

Hadar Y (2013) Sources for lignocellulosic raw materials for the production of ethanol. Lignocellulose conversion, Springer, pp 21–38

Harel A (2009) Noritech seaweed biotechnology Inc. Algae World Conference, Rotterdam, NL

Hayat S, Maheshwari P, Wani AS, Irfan M, Alyemeni MN, Ahmad A (2012) Comparative effect of 28 homobrassinolide and salicylic acid in the amelioration of NaCl stress in *Brassica juncea* L. Plant Physiol Biochem 53:61–68

Hayes DJ (2009) An examination of biorefining processes, catalysts and challenges. Catal Today 145(1–2):138–151

Hendriks A, Zeeman G (2009) Pretreatments to enhance the digestibility of lignocellulosic biomass. Bioresour Technol 100(1):10–18

Horwitz AA, Walter JM, Schubert MG, Kung SH, Hawkins K, Platt DM et al (2015) Efficient multiplexed integration of synergistic alleles and metabolic pathways in yeasts via CRISPR-Cas. Cell Syst 1(1):88–96

Isikgor FH, Becer CR (2015) Lignocellulosic biomass: a sustainable platform for the production of bio-based chemicals and polymers. Polym Chem 6(25):4497–4559

Jhunjhunwala S, van Zelm MC, Peak MM, Cutchin S, Riblet R, van Dongen JJ et al (2008) The 3D structure of the immunoglobulin heavy-chain locus: implications for long-range genomic interactions. Cell 133(2):265–279

Jönsson LJ, Martín C (2016) Pretreatment of lignocellulose: formation of inhibitory by-products and strategies for minimizing their effects. Bioresour Technol 199:103–112

Kádár Z, Szengyel Z, Réczey K (2004) Simultaneous saccharification and fermentation (SSF) of industrial wastes for the production of ethanol. Ind Crops Prod 20(1):103–110

Kalogeris E, Christakopoulos P, Katapodis P, Alexiou A, Vlachou S, Kekos D et al (2003) Production and characterization of cellulolytic enzymes from the thermophilic fungus *Thermoascus aurantiacus* under solid state cultivation of agricultural wastes. Process Biochem 38(7):1099–1104

Kang Q, Appels L, Tan T, Dewil R (2014) Bioethanol from lignocellulosic biomass: current findings determine research priorities. Sci World J

Kim S, Baek SH, Lee K, Hahn JS (2013) Cellulosic ethanol production using a yeast consortium displaying a minicellulosome and β-glucosidase. Microb Cell Fact 12(1):14

Kim S, Lee MJ, Kim H, Kang M, Kim JS (2011) Preassembled zinc-finger arrays for rapid construction of ZFNs. Nat Methods 8(1):7

Kleinstiver BP, Prew MS, Tsai SQ, Nguyen NT, Topkar VV, Zheng Z et al (2015) Broadening the targeting range of *Staphylococcus aureus* CRISPR-Cas9 by modifying PAM recognition. Nat Biotechnol 33(12):1293–1298

Klinke HB, Thomsen A, Ahring BK (2004) Inhibition of ethanol-producing yeast and bacteria by degradation products produced during pre-treatment of biomass. Appl Microbiol Biotechnol 66(1):10–26

Kumar A, Usmani Z, Kumar V (2017) Biochar and flyash inoculated with plant growth promoting rhizobacteria act as potential biofertilizer for luxuriant growth and yield of tomato plant. J Environ Manage 190:20–27
Larsson S, Palmqvist E, Hahn-Hägerdal B, Tengborg C, Stenberg K, Zacchi G et al (1999) The generation of fermentation inhibitors during dilute acid hydrolysis of softwood. Enzyme Microb Technol 24(3–4):151–159
Lewis JA, Elkon IM, McGee MA, Higbee AJ, Gasch AP (2010) Exploiting natural variation in *Saccharomyces cerevisiae* to identify genes for increased ethanol resistance. Genetics 186(4):1197–1205
Limayem A, Ricke SC (2012) Lignocellulosic biomass for bioethanol production: current perspectives, potential issues and future prospects. Prog Energy Combust Sci 38(4):449–467
Liu ZL, Weber SA, Cotta MA, Li SZ (2012) A new β-glucosidase producing yeast for lower-cost cellulosic ethanol production from xylose-extracted corncob residues by simultaneous saccharification and fermentation. Bioresour Technol 104:410–416
López M, Nichols N, Dien B, Moreno J, Bothast R (2004) Isolation of microorganisms for biological detoxification of lignocellulosic hydrolysates. Appl Microbiol Biotechnol 64(1):125–131
Lynd LR, Weimer PJ, Van Zyl WH, Pretorius IS (2002) Microbial cellulose utilization: fundamentals and biotechnology. Microbiol Mol Biol Rev 66(3):506–577
Ma M, Liu LZ (2010a) Quantitative transcription dynamic analysis reveals candidate genes and key regulators for ethanol tolerance in *Saccharomyces cerevisiae.* BMC Microbiol 10(1):169
Ma M, Liu ZL (2010b) Mechanisms of ethanol tolerance in *Saccharomyces cerevisiae.* Appl Microbiol Biotechnol 87(3):829–845
Mena-Violante HG, Olalde-Portugal V (2007) Alteration of tomato fruit quality by root inoculation with plant growth-promoting rhizobacteria (PGPR): *Bacillus subtilis* BEB-13bs. Sci Hort 113(1):103–106
Meng L, Zhang A, Wang F, Han X, Wang D, Li S (2015) Arbuscular mycorrhizal fungi and rhizobium facilitate nitrogen uptake and transfer in soybean/maize intercropping system. Front Plant Sci 6:339
Menon V, Rao M (2012) Trends in bioconversion of lignocellulose: Biofuels, platform chemicals & biorefinery concept. Prog Energy Combust Sci 38(4):522–550
Miller JC, Tan S, Qiao G, Barlow KA, Wang J, Xia DF et al (2011) A tale nuclease architecture for efficient genome editing. Nat Biotechnol 29(2):143
Monte J, Carvalho W, Milagres A (2010) Use of a mixture of thermophilic enzymes produced by the fungus *Thermoascus aurantiacus* to enhance the enzymatic hydrolysis of the sugarcane bagasse cellulose. Am J Agri Biol Sci 5(4):468–476
Neves MAD, Kimura T, Shimizu N, Nakajima N (2007) State of the art and future trends of bioethanol production. Dyn Biochem Proc Biotechnol Mol Biol 29:1–14
Nigam J (2002) Bioconversion of water-hyacinth (*Eichhornia crassipes*) hemicellulose acid hydrolysate to motor fuel ethanol by xylose–fermenting yeast. J Biotechnol 97(2):107–116
Ocreto MB (2013) Delignification of lignocellulosic biomass for bioethanol production. USM R&D J 21(1):1–20
Öhgren K, Bura R, Lesnicki G, Saddler J, Zacchi G (2007) A comparison between simultaneous saccharification and fermentation and separate hydrolysis and fermentation using steam-pretreated corn stover. Process Biochem 42(5):834–839
Öhgren K, Bura R, Saddler J, Zacchi G (2007) Effect of hemicellulose and lignin removal on enzymatic hydrolysis of steam pretreated corn stover. Bioresour Technol 98(13):2503–2510
World Health Organization (2015) Food and agriculture organization of the United Nations. Probiotics in food: Health and nutritional properties and guidelines for evaluation. Fao Food and Nutrition, Rome
Palmqvist E, Hahn-Hägerdal B (2000) Fermentation of lignocellulosic hydrolysates. II: Inhibitors and mechanisms of inhibition. Bioresour Technol 74(1):25–33

Panagiotou G, Kekos D, Macris BJ, Christakopoulos P (2003) Production of cellulolytic and xylanolytic enzymes by *Fusarium oxysporum* grown on corn stover in solid state fermentation. Ind Crops Prod 18(1):37–45

Park Y, Kang S, Lee J, Hong S, Kim S (2002) Xylanase production in solid state fermentation by *Aspergillus niger* mutant using statistical experimental designs. Appl Microbiol Biotechnol 58(6):761–766

Pham LJ, Halos SC (1990) Intergeneric protoplast fusion of *Trichoderma reesei* RUT C-30 and *Penicillium funiculosum* Thom MG-171 for improved cellulase production. Ann N Y Acad Sci 613(1):575–581

Qureshi N, Dien B, Liu S, Saha B, Hector R, Cotta M et al (2012) Genetically engineered *Escherichia coli* FBR5: Part I. Comparison of high cell density bioreactors for enhanced ethanol production from xylose. Biotechnol Prog 28(5):1167–1178

Rabemanolontsoa H, Saka S (2016) Various pretreatments of lignocellulosics. Bioresour Technol 199:83–91

Rouches E, Herpoël-Gimbert I, Steyer J, Carrere H (2016) Improvement of anaerobic degradation by white-rot fungi pretreatment of lignocellulosic biomass: a review. Renew Sustain Energy Rev 59:179–198

Rutherford BJ, Dahl RH, Price RE, Szmidt HL, Benke PI, Mukhopadhyay A et al (2010) Functional genomic study of exogenous n-butanol stress in *Escherichia coli*. Appl Environ Microbiol 76(6):1935–1945

Saha BC, Iten LB, Cotta MA, Wu YV (2005) Dilute acid pretreatment, enzymatic saccharification and fermentation of wheat straw to ethanol. Process Biochem 40(12):3693–3700

Saini JK, Saini R, Tewari L (2015) Lignocellulosic agriculture wastes as biomass feedstocks for second-generation bioethanol production: concepts and recent developments. 3 Biotech 5(4):337–353

Sanchez OJ, Cardona CA (2008) Trends in biotechnological production of fuel ethanol from different feedstocks. Bioresour Technol 99(13):5270–5295

Saratale GD, Saratale RG, Lo YC, Chang JS (2010) Multicomponent cellulase production by *Cellulomonas biazotea* NCIM-2550 and its applications for cellulosic biohydrogen production. Biotechnol Prog 26(2):406–416

Sarkar N, Ghosh SK, Bannerjee S, Aikat K (2012) Bioethanol production from agricultural wastes: an overview. Renew Energy 37(1):19–27

Saxena J, Saini A, Ravi I, Chandra S, Garg V (2015) Consortium of phosphate-solubilizing bacteria and fungi for promotion of growth and yield of chickpea (*Cicer arietinum*). J Crop Improv 29(3):353–369

Schirmer A, Rude MA, Li X, Popova E, Del Cardayre SB (2010) Microbial biosynthesis of alkanes. Science 329(5991):559–562

Shin SE, Lim JM, Koh HG, Kim EK, Kang NK, Jeon S et al (2016) CRISPR/Cas9-induced knockout and knock-in mutations in *Chlamydomonas reinhardtii*. Sci Rep 6:27810

Silva DD, Arruda PV, Dussán KJ, Felipe MG (2014) Adaptation of *Scheffersomyces stipitis* cells as a strategy to the improvement of ethanol production from sugarcane bagasse hemicellulosic hydrolysate. Chem Eng 38

Silva DDV, Dussán KJ, Hernández V, da Silva SS, Cardona CA, de Almeida Felipe MdG (2016) Effect of volumetric oxygen transfer coefficient (kLa) on ethanol production performance by *Scheffersomyces stipitis* on hemicellulosic sugarcane bagasse hydrolysate. Biochem Eng J 112:249–257

Singh P, Suman A, Tiwari P, Arya N, Gaur A, Shrivastava A (2008) Biological pretreatment of sugarcane trash for its conversion to fermentable sugars. World J Microbiol Biotechnol 24(5):667–673

Singhania RR, Patel AK, Pandey A, Ganansounou E (2017) Genetic modification: a tool for enhancing beta-glucosidase production for biofuel application. Bioresour Technol

Slininger PJ, Shea-Andersh MA, Thompson SR, Dien BS, Kurtzman CP, Balan V et al (2015) Evolved strains of *Scheffersomyces stipitis* achieving high ethanol productivity on acid-and base-pretreated biomass hydrolyzate at high solids loading. Biotechnol Biofuels 8(1):60
Smith J, Grizot S, Arnould S, Duclert A, Epinat JC, Chames P et al (2006) A combinatorial approach to create artificial homing endonucleases cleaving chosen sequences. Nucleic Acids Res 34(22):e149–e149
Stefanidis SD, Kalogiannis KG, Iliopoulou EF, Michailof CM, Pilavachi PA, Lappas AA (2014) A study of lignocellulosic biomass pyrolysis via the pyrolysis of cellulose, hemicellulose and lignin. J Anal Appl Pyrolysis 105:143–150
Sun Y, Cheng J (2002) Hydrolysis of lignocellulosic materials for ethanol production: a review. Bioresour Technol 83(1):1–11
Sunna A, Gibbs MD, Chin CW, Nelson PJ, Bergquist PL (2000) A gene encoding a novel multidomain β-1, 4-mannanase from *Caldibacillus cellulovorans* and action of the recombinant enzyme on kraft pulp. Appl Environ Microbiol 66(2):664–670
Szczodrak J, Fiedurek J (1996) Technology for conversion of lignocellulosic biomass to ethanol. Biomass Bioenergy 10(5–6):367–375
Talebnia F, Karakashev D, Angelidaki I (2010) Production of bioethanol from wheat straw: an overview on pretreatment, hydrolysis and fermentation. Bioresour Technol 101(13):4744–4753
Ulaganathan K, Goud S, Reddy M, Kayalvili U (2017) Genome engineering for breaking barriers in lignocellulosic bioethanol production. Renew Sustain Energy Rev 74:1080–1107
Voloshin RA, Rodionova MV, Zharmukhamedov SK, Veziroglu TN, Allakhverdiev SI (2016) Biofuel production from plant and algal biomass. Int J Hydrog Energy 41(39):17257–17273
Wagaba H, Patil BL, Mukasa S, Alicai T, Fauquet CM, Taylor NJ (2016) Artificial microRNA-derived resistance to Cassava brown streak disease. J Virol Methods 231:38–43
Werther J, Saenger M, Hartge EU, Ogada T, Siagi Z (2000) Combustion of agricultural residues. Prog Energy Combust Sci 26(1):1–27
Wiedenheft B, Sternberg SH, Doudna JA (2012) RNA-guided genetic silencing systems in bacteria and archaea. Nature 482(7385):331–338
Wu WH, Hung WC, Lo KY, Chen YH, Wan HP, Cheng KC (2016) Bioethanol production from taro waste using thermo-tolerant yeast *Kluyveromyces marxianus* K21. Bioresour Technol 201:27–32
Xu RF, Li H, Qin RY, Li J, Qiu CH, Yang YC et al (2015) Generation of inheritable and "transgene clean" targeted genome-modified rice in later generations using the CRISPR/Cas9 system. Sci Rep 5:11491
Zabed H, Sahu J, Boyce A, Faruq G (2016) Fuel ethanol production from lignocellulosic biomass: an overview on feedstocks and technological approaches. Renew Sustain Energy Rev 66:751–774
Zhao XQ, Zi LH, Bai FW, Lin HL, Hao XM, Yue GJ et al (2011) Bioethanol from lignocellulosic biomass. In: Biotechnology in China III: Biofuels bioenergy, Springer, pp 25–51
Zheng Y, Pan Z, Zhang R (2009) Overview of biomass pretreatment for cellulosic ethanol production. Int J Agric Biol Eng 2(3):51–68
Zhou C, Zhu L, Ma Z, Wang J (2017) *Bacillus amyloliquefaciens* SAY09 increases cadmium resistance in Plants by activation of auxin-mediated signaling pathways. Genes 8(7):173
Zhou S, Yomano L, Shanmugam K, Ingram L (2005) Fermentation of 10%(w/v) sugar to D (−)-lactate by engineered *Escherichia coli*. Biotechnol Lett 27(23–24):1891–1896
Zhu J, Pan X (2010) Woody biomass pretreatment for cellulosic ethanol production: technology and energy consumption evaluation. Bioresour Technol 101(13):4992–5002
Zhu J, Pan X, Wang G, Gleisner R (2009) Sulfite pretreatment (SPORL) for robust enzymatic saccharification of spruce and red pine. Bioresour Technol 100(8):2411–2418
Zhu J, Wang G, Pan X, Gleisner R (2009) Specific surface to evaluate the efficiencies of milling and pretreatment of wood for enzymatic saccharification. Chem Eng Sci 64(3):474–485

Chapter 6
Impact of CO_2 Discharge from Distilleries on Climate Changes: Key Facts

Aneela Sabir, Humaira Idrees, Muhammad Shafiq, Muhammad Taqi Zahid Butt, Karl I. Jacob, and Muhammad Arshad

Abstract Climate change due to the rapid release of carbon dioxide from distilleries may result in creating many drastic changes in the atmosphere such as floods, severe change in temperature that greatly affect human health. CO_2 is an important Greenhouse gas is useful for the atmosphere when present at a concentration of up to 0.03% or 390 ppm, but due to rapid industrialization the exceeded level of CO_2 in the atmosphere trapped the harmful radiation from the Sun and shutdown the Greenhouse window. The anthropogenic effects of CO_2 on climate can be mitigated through various methods, among them, sequestration through geological media is more economical, and this technology has been developed for hydrocarbon production and exploration. Different geological methods such as Oil displacement in reservoirs, use of methane for sequestration of CO_2 in coal beds, storage in depleted hydrocarbon reservoirs, and sequestration of CO_2 through membrane technology (Nano-filtration and Reverse Osmosis) are the other useful methods.

Keywords Distilleries effluent · Greenhouse gas · Carbon dioxide · Climate change · Sequestration

6.1 Introduction

Due to the rapid growth of industrialization (distilleries or spent wash) and urbanization, there is release of different types of pollutants in the environment (Sabir et al. 2016a, b, c, d). Furthermore, there is concern about the safety of nuclear reactors, radioactive waste disposal, distilleries discharge, and acid rain. Today, global consideration has stressed on the anthropogenic emissions of hazardous gases to the atmosphere, the ozone hole, and increase of greenhouse effect along with the above-mentioned anthropogenic actions include the main source of environmental change.

A. Sabir (✉) · H. Idrees · M. Shafiq · M. T. Z. Butt · K. I. Jacob
Department of Chemical Engineering, University of the Punjab Lahore, Lahore, Pakistan
e-mail: aneela.pet.ceet@pu.edu.pk

M. Arshad
Jhang-Campus, University of Veterinary and Animal Sciences Lahore, Lahore, Pakistan

M. Arshad (ed.), *Sustainable Ethanol and Climate Change*,
https://doi.org/10.1007/978-3-030-59280-6_6

Hazardous chemicals are discharged directly or indirectly into the environment and result in an increase in the concentration of greenhouse gases in the environment. Presently, the greenhouse effect grasps the central phase with respect to transnational efforts to express and apply adequate environmental strategies (Ribeiro and Rosa 2017).

Distillery waste has a very high pollution load, with a Biochemical Oxygen Demand (BOD) value of 40,000 to 60,000 mg l^{-1} and Chemical Oxygen Demand (COD) value of 60,000 to 90,000 mg l^{-1}. Distilleries effluents with this value of COD and BOD are highly acidic having a pH value of 4.5 (Viswanathan and Kumar 1991). And distilleries discharge also released CO_2 into environment which is the main greenhouse gas (Rubin 2005; IPCC 2007). Differences of volcanic vaporizers and possibly solar luminosity appear to be the primary reason for variations of increasing temperature. It is shown in this chapter that how anthropogenic carbon dioxide heating arise the climate variability (Hansen et al. 2017). In June 2016, according to the data measured from an observatory at Mauna Loa in Hawaii, the atmospheric carbon dioxide (CO_2) amount increased up to 400 ppm. As 450 ppm has been set as a target level for stabilizing the climate at 2 °C. Between many GHGs, carbon dioxide is considered as a primary GHG as it accounts for the largest portion (77%) of the global GHG emissions (Roh et al. 2016).

By description, an air pollutant is any material which may harm animals, vegetation animals, vegetation humans, or material. As far as humans are concerned an air contaminant may cause or contribute to an increase in serious illness or mortality (Kampa and Castanas 2008). The disorders caused due to the climate change in human health are depression, posttraumatic stress, stomach disorder, child abuse, suicide, mortality rate, heat stroke, mental disorder, etc.

And in most of the countries of the world people are facing vector born and nutritional disorders (Pascual et al. 2006).

Due to increasing attention on sustainable and social accountability, administrations have developed better goals more than economic profit, such as obligation to social and environmental results (Kiron et al. 2012). Moreover 2800 global companies discovered that 70% of these organizations contain sustainability as the main issue in their intentional plans and outlines.

The UN General Assembly presented "the 2030 Agenda for sustainable development" having 169 targets and 17 sustainable development goals. The 17 goals are combined in three dimensions of sustainable development: social, environmental, and economic. They are recognized on what is identified as the five Ps: "planet, Prosperity, people, peace and partnership." With regard to "people" and "prosperity," the task of the SDGs is to develop the proper setting and develop a proper condition that improves the development of sustained economic growth. From the business point of view, SDG's is to develop the innovative, sustainable and people-oriented. The accomplishment of SDGs requires a planned process: the public and private sectors, multi-national enterprises, governments, non-governmental, and individual's philanthropic organizations (Chams and García-Bland6n 2019).

So, the anthropogenic effects of CO_2 on climate can be eased through various methods; among them, sequestration through geological media is more economical and this technology has been developed for hydrocarbon production and exploration. Different geological methods such as oil displacement in reservoirs, use of methane for sequestration of CO_2 in coal beds, storage in depleted hydrocarbon reservoirs, and sequestration of CO_2 through membrane technology (Nano filtration and Reverse Osmosis) is also useful (Sabir et al. 2014, 2015, 2016a, b, c, d).

6.2 Introduction of Distilleries and Spent Wash

Distillery spent wash means the effluent produced due to alcohol distilleries. On an average, 8–15 L of effluent is generated for every liter of alcohol produced in the alcohol distilleries which are extensively growing due to widespread industrial applications of alcohol such as in food, pharmaceuticals, and perfumery. It is also used as a substitute fuel. In India, there are 319 distilleries, producing 3.25 billion liters of alcohol and generating 40.4 billion liters of wastewater annually. According to Ministry of Environment and Forests (MoEF), alcohol distilleries are listed at the top in the "Red Category" industries. Alcohol production in distilleries consists of four main steps viz., fermentation, distillation, feed preparation and packaging. Ethanol can be arranged from various biomass materials but the potential for their use as feedstock depends on the availability, cost, carbohydrate contents, and the ease by which they can be converted to alcohol. About 61% of the world's ethanol production is carried out from sugar crops. Most of the distilleries exclusively use cane syrup as raw material for fermentation. Sugar syrup is concentrated and then it is diluted with ammonium Sulphate, urea, and any other source of nitrogen. Distilleries effluents having high concentrations of Biological oxygen demand (BOD) and Chemical oxygen demand (COD) amounts of spent wash due to the occurrence of huge organic substances. It includes lignin, carbohydrates proteins, waxes, and reduced sugars (Mohana and Acharya et al. 2009).

Process flow diagram of CO_2 release from distilleries and reverse osmosis (RO) and nanofiltartion (NF) (pressure of 0–7 bar) Pilot Plants are used for the treatment of Distilleries effluents. Moreover, the hybrid RO and NF was also used for the treatment of industrial effluents (Figs. 6.1 and 6.2).

6.3 Physics and Chemistry of Climate

Physics and chemistry of our earth are basically explained by climate. Climate can be defined as long-term average weather patterns for a specific region. Change in climate is actually long-lasting state of atmosphere for several decades. According to literature, climate change is also explained as the combination of human-induced or natural climate changes or may be only due to human persuaded variations (George Luber).

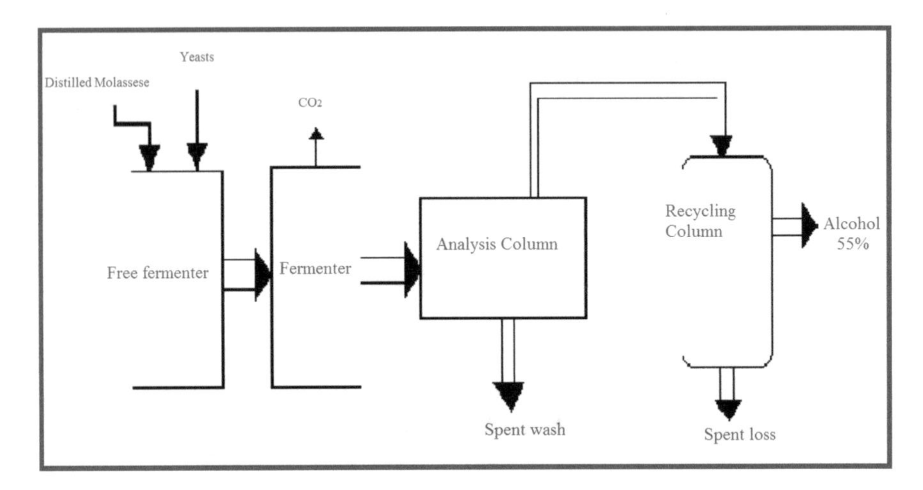

Fig. 6.1 Release of CO_2 from distilleries

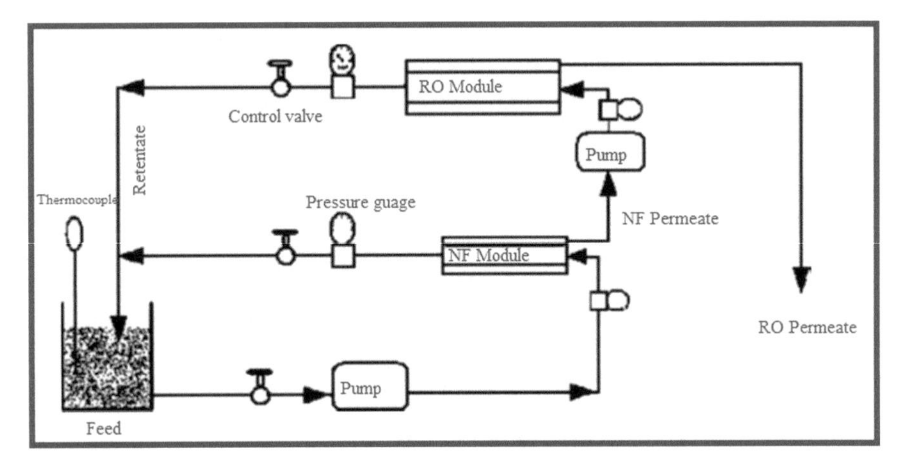

Fig. 6.2 Pilot plant used for treatment of effluents released from distilleries

6.4 Different Layers of Atmosphere:

6.4.1 Four Main Layers of Atmosphere

- **Thermosphere**
- **Mesosphere**
- **Stratosphere**
- **Troposphere**

See Fig. 6.3.

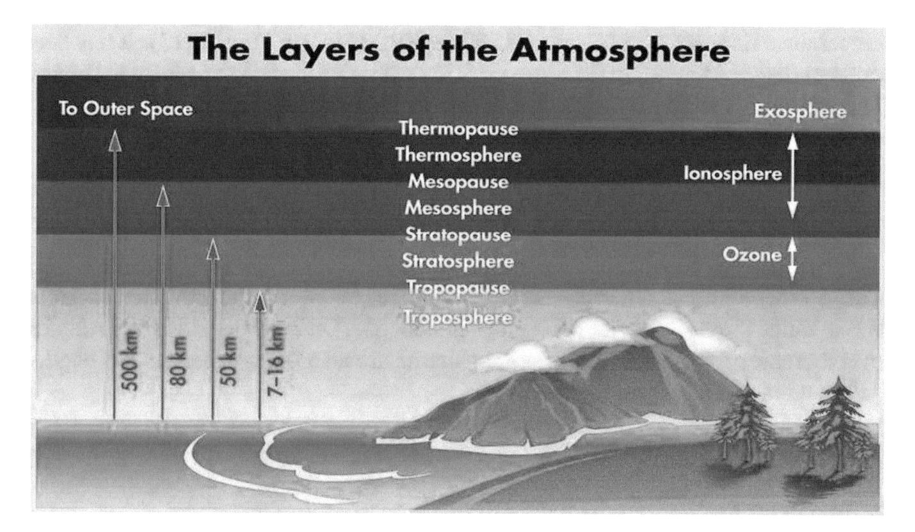

Fig. 6.3 Layers of atmosphere

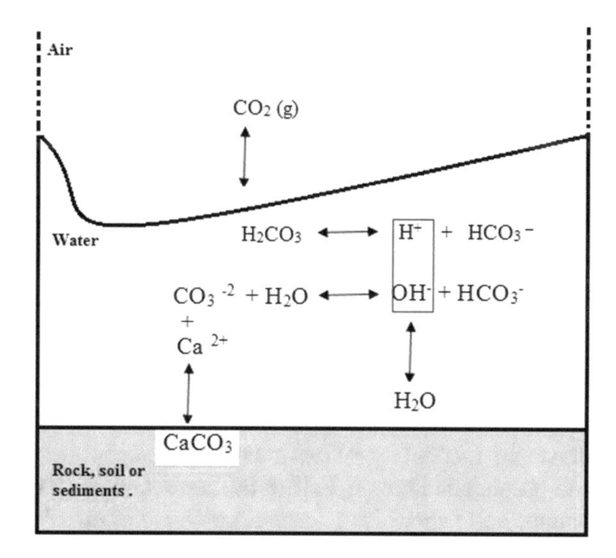

Fig. 6.4 Reaction of carbonate formation in Oceans

6.5 Climate Change Due to Earth Temperature Variation

Earth temperature near the poles of earth is low, and at equator, temperature rises. In tropical region, weather is mostly warmer because of elevated temperature, the reason being that the maximum amount of solar radiations reaches this area. The difference in temperature generates a large amount of atmospheric rotation, which distributes energy from tropical to high-latitude regions. Since early twentieth century, earth temperature rises by 0.85 °C, from 1880 to 2012 (Stocker et al. 2013). It has been reported that an increase in temperature is almost twice in each decade up to 0.13 °C (Hansen et al. 2010).

6.5.1 Effect of Climate on Antarctic and Arctic Region

Since 1978, data collected from satellite allowed us to observe the reduction of Arctic sea ice ranges from 3.5 to 4.1 percentage per decade. On the other side, Antarctic ice level increases from 1.2 to 1.8%; wind patterns are also responsible for this change (Sillmann et al. 2013).

6.5.2 Evidences of Climate Variation

Catastrophic Climate variations that earth has experienced from the last few decades are becoming a source of interest in the scientific community. There are many uncertainties regarding the impact of human activities such as industrial uprising, increasing the atmospheric concentrations of gases. Atmosphere looks like a giant basin for absorption of a large amount of distillery gases. But the atmospheric layer around the earth is very thin as its capacity to absorb these poisonous gases reduces day by day (Edward 1997, p. 209). The gases present in large amount in our atmosphere are Argon (0.93%), Nitrogen (78.09%), oxygen (20.95%), and in trace amount, Chlorofluorocarbons (CFCs) Carbon monoxide (CO), Nitrogen oxide (NO_X), methane (CH_4) ozone, and Carbon dioxide (CO_2).

6.6 Distilleries Discharge into Oceans

The main uncertainty related to climate changes are the resources of carbon (plants and oceans) and mostly half of the reservoirs of CO_2 on land are basically plants, humus, and vegetables (Corinne Le Quéré et al. 2009). Oceans are considered as the major reservoirs having 60 times greater carbon dioxide as compared to the atmospheric (Sawyer 1972).

- First, the most important resource of CO_2 is the Distilleries effluents. Manmade CO_2 increase exponentially. Bolin et.al. proposed that the concentration of CO_2 in future will be exceeded 400 p.p.m
- Second, most reservoirs are oceans, the top-most layer of the sea having the same amount of CO_2 as the atmosphere. The reason behind the same concentration is that the decay of ^{14}C which are produced in atmosphere reduced to ^{12}C, proposed that transfer of CO_2 from atmosphere to the top layer of ocean is very slow process

required almost 5 to 10 years. Centuries are required to maintain the equilibrium of carbon dioxide in the ocean (Corinne Le Quéré et al. 2009).

6.7 Effect of Distilleries Effluent is the Evidence for Climate Change

Industrial uprising produced mostly harmful effluents having toxic threats to the different component of environment, air, soil characteristics as the chemicals leache to the groundwater. Disposal of wastes from industries is becoming a threat for environment all over the world. Groundwater having organic materials and different poisonous gases (CO, Cl_2, NO_2 and CO_2) are present which creates major pollution threats (Farid Ansari 2010). Huge volumes of effluent are often disposed on nearby areas of land, which causes worse leaching in land. Most of the industries discharge their effluent on the land openly and sometimes effluents are used for irrigation purpose. These improper disposals of effluents introduced toxic metal on the surface of the soil. On becoming the part of soil, these effluents undergo the process of oxidation, biological decay, dilution and filtration, etc. The movement of distilleries effluents along the freshwater source when it penetrates the ground produces serious complexation process, both in freshwater source and in the ground. The reason for complexation reaction is organic, inorganic legends, metals and different types of toxic gases (CO, Cl_2, NO_2 and CO_2) present in distilleries discharge. The mobility of the metals in soil is also influenced by soil cation exchange capacity, pH, soil redox potential, and organic matter.

6.8 Comparison Between Distilleries Discharge of Oceans and Atmospheric CO_2

Kanwisher. Et al. proposed that only a small amount of CO_2 is available as a dissolved CO_2 in the oceans. The presence of magnesium and Sodium carbonate in the ocean act as a buffering solution for CO_2. In short, the increased level of CO_2 in ocean up to 0.6% is responsible to increase the partial pressure of atmospheric CO_2 level up to 10%. Increase in partial pressure of CO_2 in ocean and atmosphere increases the concentration of CO_2 up to 20%. A large process is required to deposit the ocean's CO_2 to Carbonate (as shown in Figs. 6.4 and 6.5) as compared to the Industrially produced CO_2. Confusion between the oceanic and atmospheric CO_2 is that at low temperature the absorption of CO_2 in the oceans is greater as compared to the oceans present at high altitude. In most of the case, the oceans have colder top-most layer, due to which it can absorb the maximum amount of atmospheric CO_2 but the lower layer having lesser CO_2, is the reason due to which some ocean is having different composition of CO_2 in it. This difference in layers is also acting as a barrier for dissolving of the upper and lower layer. Most of the polar region oceans do not have

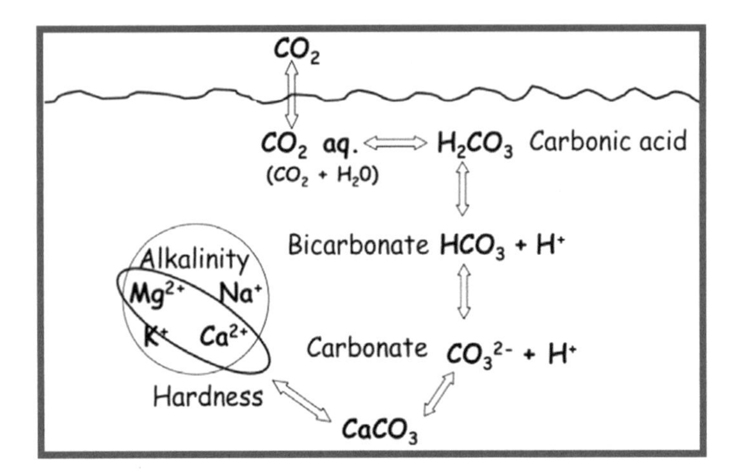

Fig. 6.5 Carbonate formation

this barrier due to which the balance between the ocean's and atmospheric CO_2 is maintained (Sawyer 1972) (Fig. 6.5).

6.9 Evidences of Climate Change Due to Distilleries Discharge

6.9.1 CO_2 as a Greenhouse Gas and the Main Cause of Global Warming Effect

Earth temperature is maintained by radiations that come directly from the Sun. Some of these radiations reflected back when they hit the clouds and most of them become the part of atmosphere, and heat it up along with oceans. Most of the radiations which reflected back have longer wavelength, because they have low temperature as compared to the Sun, and these reflected radiations are the source of rain. Radiations having short wavelength come to the earth, Greenhouse gases including water vapors absorbed these radiations and are responsible for drastic change of earth atmospheric temperature (Sawyer 1972).

Greenhouse effect is actually the natural phenomenon which allows the survival of life on earth. Atmospheric main constituents are oxygen and nitrogen which cannot absorb and emit the radiation. Radiation come from Sun with a frequency of 400–750 nm, hit the earth surface and emit again with lower energy into infrared region (4000–50,000 nm) back into space. But the other gases like methane, CFC's, Nitrous oxide, water vapors, and carbon dioxide are absorbed in the infrared region. These gases mainly CO_2 is Greenhouse gas and also the major cause of Global warming effect. The exceeded level of CO_2 trapped the harmful radiations in the atmosphere (radiations not bounced back to space) due to which temperature of Earth increases day by day or in short Greenhouse windows closed. So, it is evidenced that this trace amount of gases has a significant role on earth's atmosphere.

Natural significance of Greenhouse effect is to keep the earth temperature above 20 °C, and it is possible only when the atmosphere of the earth is composed of only O_2 and N_2.

The fact can be justified by the example of planet Venus, having CO_2 about 96% and temperature is almost 500 °C.

6.10 6.9 CO_2 Responsible for the Climate Variations

Globally 80% of the CO_2 emitted from industrial and transportation activities and the remaining 20% of CO_2 comes from biomass combustion and deforestation. Forestation is beneficial for CO_2 storage, they stored about 100 tons of CO_2. But in the last few years it was observed that half of the forests was destroyed completely. CO_2 enhanced the greenhouse effect about 64% and the energy sources are such as fossil fuels (oil, gas, and petroleum) donates 85% of the anthropogenic effect on climate to increase the concentration of CO_2 in the environment (Jepma and Munasinghe 1998, p. 331; Intergovernmental Panel on Climate Change 1991, p. 272).

6.10.1 *Direct and In-Direct Effect of CO_2*

Direct effect of CO_2 might be ignored (may be affects the plant and seed growth), but the indirect effect of CO_2 on the mankind is very clear as Greenhouse effect is the major example of indirect effect of CO_2.

6.10.2 *Human Activities as a Source of CO_2*

The main cause of these gases is human activities such as burning of fossil fuels and distilleries effluents. Distilleries effluents are continuously entering the pollutants in the atmosphere and affect the sources of freshwater, as well as these gases enter into the atmosphere. Scientist who supports this fact that CO_2 is the main cause of Global warming proposed that the change in climate in the last few years is very abrupt. And the main cause of this abrupt change in climate is due to continuous emission of CO_2. Recent report of Intergovernmental Panel on climate change justified that human activities are the main reason for climate change (Fig. 6.6).

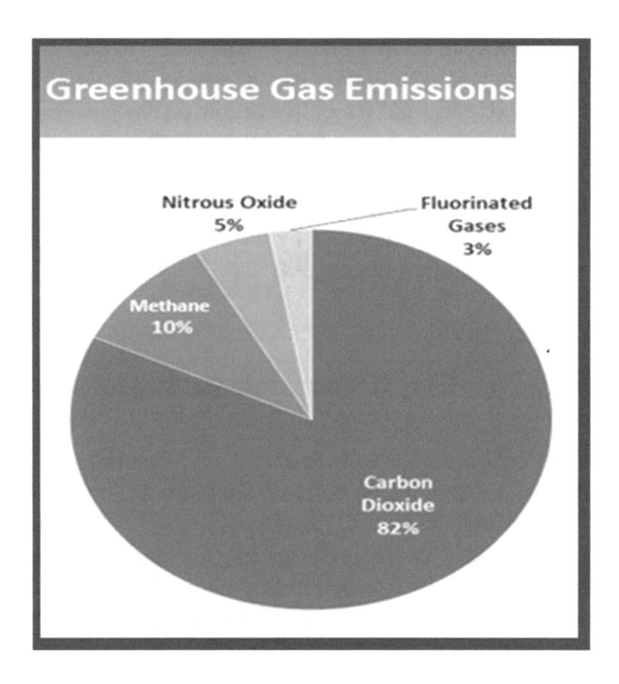

Fig. 6.6 Emission of GreenHouse gases

6.10.3 Effect of CO_2 on Global Warming

6.10.3.1 Effect of Climate Change on Hydrological Cycle

Rainfall, precipitation, and any other form of solid that fall on earth occurs when the atmosphere absorbs the water from Earth. Different geographical processes generate the variations in annual precipitation in the past century. Hot weather then normally enhance the hydropower generations, wildfires, reduced agriculture yield, different diseases, etc. These changes also affect the social and economic functions of society. Furthermore, higher temperature leads to more rainfall events then snowfall, these variations occur from North America to South Asia and also decrease in the volume, mass, and area around the globe (Brown 2009).

6.10.3.2 Effect on Power Plants Due Water Supply Shortage

The aim of scientists under this heading is to study the strategies for an ecological management of water on a regional scale by considering the socio-economic conditions and global environmental conditions. Due to increase in the duration of dry and hot weather there is shortage of water for power plants operation. Due to estimation of the scientists, the various power plants had to be throttled in summer due to hot and dry climate (Intergovernmental Panel on Climate Change 1991; Zebisch et al. 2005; Jepma 1998, p. 331.). Power plants water demands are conquered by their need for cooling processes. Two types of processes can be used Once-through and Closed Circuits (Edward 1997; Wurbs 2005; IPCC 2007; Fujii and Iwata et al. 2017). Less amount of water consumes in Once-through systems but actually required large quantities in the cooling process. Smaller amount of water is required in Closed-circuit systems also but most of the water is lost in evaporation process (Gleick 1994). However, changing the cooling system from once-through to closed circuit could expressively increase the water feeding (Feeley et al. 2008). Freeley et al. studied the scenario including the process of different trends of energy demand and cooling systems in power plants. In 2030, water demand for power plants reduces to 30% or remain still at the present level. Effect on power plant due to climate change is studied by EPRI, Hurd and Harrod, and Kirshen et al. (EPRI 1995; Hurd and Harrod 2001; Kirshen et al. 2008). The investigation by Hurd and Harrod showed very small to distinct losses as a result of climate change, depending on the region analyses. EPRI proposed the potential costs related to climate change are minor for existing power plants. However, this assessment was based on past data rather than on climate change situations. These investigations suggested that the water scarcities and limits on heat ejections would lead to economic losses for power plants. Climate change affects the water supply and the water accessibility for power plants. Coupled simulation was used, and we can estimate whether the variation capacities of power plants to altering socio-economic and climate circumstances are enough or whether huge economic costs may occur (Werner and Gerstengarbe 1997).

The climate change and socio-economic effect on power plants are taken into account by considering:

- Specific factors by using cooling towers.
- Technological changes and Energy demands can be developed
- Cost of the water shortage
- Availability of the water at power plants.

Most of the water reserved by power plants is used in the cooling process. In order to recognize the use of water in power plants, it is essential to explain the degeneracy of waste heat in power plants. The amount of waste heat is affected by different parameters such as power plant technology, efficiency and load of operation, (Koch and Vögele 2009). The cooling systems used mostly in power plants to scatter waste heat are illustrated in Fig. 6.7 (Fig. 6.8).

6.10.3.3 *Elevated CO_2 Concentration Cause Flooding*

From 1870, the sea level of earth globally increases by 8 inches due to thermal expansion of water and melting of glaciers and polar ice caps which in turn results to flooding; flooding is a multifaceted phenomenon that can be caused by many processes, such as long-lasting snowmelt, precipitation, barrier break, ice jam, land-slide, and coastal flooding. Floods in the last decade have huge amount of material losses about 10 billion US $, economic losses increase by ten folds and death ratio

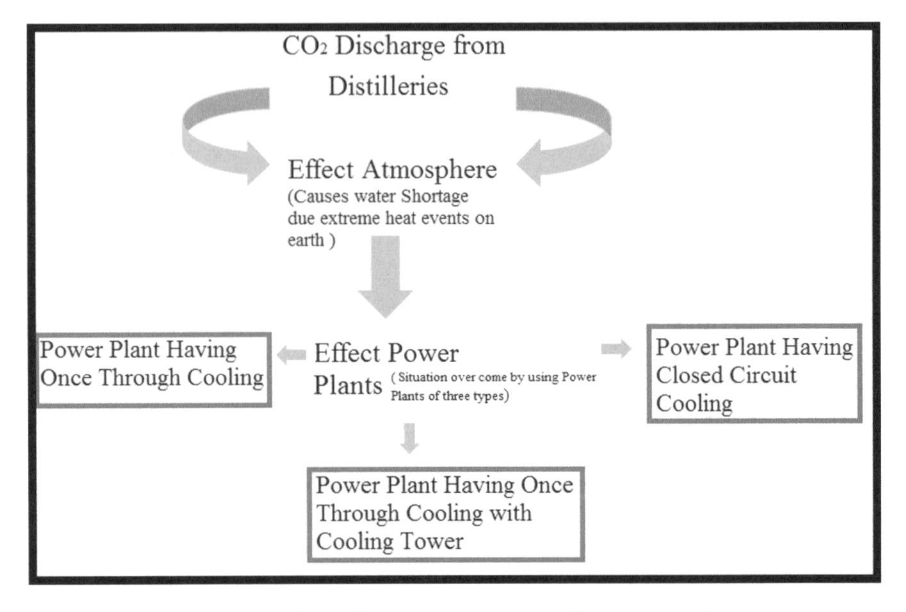

Fig. 6.7 Flow sheet of atmospheric effect on power plant and its overcome situation

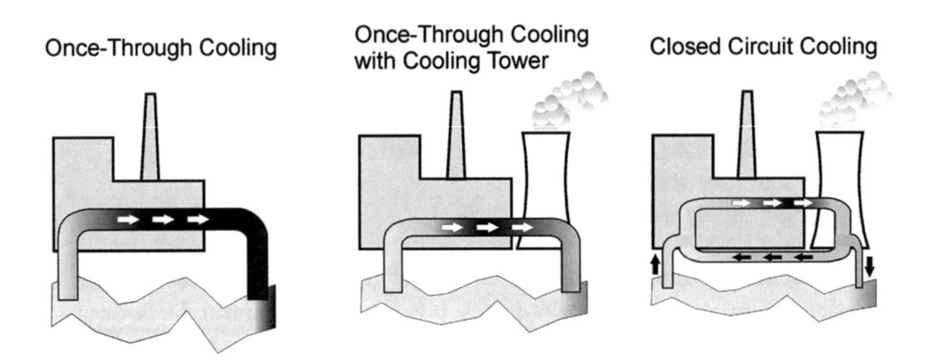

Fig. 6.8 Comparison of cooling systems for power plants

is more than 1,000 (Kanae et al. 2010). Critical floods commonly occurred in lower latitude regions, particularly in Asia (especially in china, Bangladesh, and South America) and India, Pakistan. Nearly, 70% of the country area was inundated in Bangladesh, during flood in 1998. Destructive floods are also observed in different countries, also in Europe in the year 2002 having huge amount of material loss (Kanae et al. 2010).

6.10.4 *Effect of Climate Change on Surface Water*

Global warming effect and other factors of pollution caused by exceeded amount of CO_2 alter the quality of surface water. Freshwater ecosystems are a type of social and economic status throughout the world and important for the quality of life, public health, and environmental stability of nations. Humans not only fulfill the need of food and water from aquatic ecosystem, but they also maintain the filtering quality of water, removal of contaminants, helpful in transportation, refill farmland nutrients through periodic flooding, sustained the food and habitat for wildlife, and provide entertaining opportunities. More than one-third of the Earth's freshwater that drains from land to ocean is currently used for human needs (Justic et al. 1996; Postel et al. 1996). Pollutants discharged from distilleries directly or indirectly enter into the water produces huge amount of toxicity in the water disturbed the aquatic ecosystem (Nielson and Lee 1987; Jacoby 1990; Mueller et al. 1995; Fritz 1996; McKnight et al. 1996; Watson et al. 1996; Dale 1997).

6.10.5 *Effect of Climate Change on Extreme Heat Events and Human Health*

Poverty is an autonomous risk factor for ailment associated with heat and it is linked with a decreased possibility of the medical care and lack of proper measurements of air-conditioning. These extreme heat events of the heated island are due to the emission of heat radiation during night from buildings, from industrial heat source, and temperature events of both day time and night. These heat events lead to increasing mortality rate in Chicago 1995. Disaster due to drought also affect human health and it is much more as compared to the floods, wildfire, storms, and land sliding. About 56.6% of disaster occur due to drought, 34.6% due to floods, 4.95% due to storms, 0.01% due to wildfires, land sliding generates 0.14% (George Luber; Femke Vos et al. 2009). Disorder caused due to the climate change in human health are depression, posttraumatic stress, stomach disorder, child abuse, suicide, mortality rate, heat stroke mental disorder, etc. At high altitude, the people face vulnerable health issues such as vector borne diseases and malaria. According to literature, the gender and age issues due to severe cyclone in 1991 in Bangladesh around about 140,000 people were killed and death rate of children under the age of 10 increased (Pascual et al. 2006). Most of the children under five-year age in developing countries face malaria and water-born disease such as diarrhea, and most of the diseases are caused due to lack of nutrients (Bern et al. 1993). Globally, poor children suffer from climate-sensitive diseases and in developing countries mostly diseases spread due to soil, water, and vector such as hookworm, schistosomiasis, and vectors (Hotez et al. 2008). According to an estimation about 170 million peoples in Africa will suffer from malaria in 2030 (Richardson et al. 2011; Hay et al. 2006).

6.10.6 Management of CO_2 Necessity and Technologies

According to many models, the accepted concentration of CO_2 is 450 ppm, as this concentration is stabilized the world temperature by 2 °C at the end of twenty-first century (Hare et al. 2011). Global CO_2 monitoring at Mauna Loa in Hawaii shows that the concentration of CO_2 is already above 400 ppm (Roh et al. 2016), that is, almost nearing the recognized target limit. To reduce the global carbon dioxide releases, different methods have been proposed:

- improvement of process productivities to reduce energy consumption (Kaplan 2015).
- consumption of alternative energy sources (e.g., wind power, solar energy, and geothermal) and renewable (e.g., biomass and biofuel) (Karakosta et al. 2013).
- Carbon dioxide capture, utilization and sequestration technologies (Critoph 1988).

6.10.6.1 CO_2 Capture, Sequestration, and Utilization (CCU)

Although CCU and CCS have significant impact on the stabilization of CO_2 concentration in the atmosphere, tests in their implementation, such as additional storage, cost and other related problems, structural policies, technical capabilities and environmental difficulties have not yet been sufficiently addressed (Mohana and Acharya et al. 2009; Rahman et al. 2017).

6.10.6.2 Capture Technologies

The industrial segment accounts for about 50% of the total CO_2 releases. The concentration of CO_2 in fuel gas emissions from stationary sources is relatively higher as compared to the mobile source. Therefore, avoiding the release of CO_2 from stationary sources is a feasible option to control the global CO_2 concentration. To capture CO_2 from stationary sources and the atmosphere, different mass transfer techniques may be used including absorption, adsorption, and membrane separations. Techniques are well established to capture CO_2 at different stages (post combustion, pre-combustion, and oxy-combustion) for emissions from stationary sources, i.e., mainly from chemical processes and power plants.

Pre-Combustion Capture Process

In pre-combustion H_2 and CO_2 are produced from fuel gasification followed by water gas shift reaction. Well established processes are available to separate CO_2 from the fuel gas by physical absorption using rectisol solvent and selexol (IEA 2013; IEA 2017).The main advantage of this process is the high concentration of CO_2 in the synthesis gas produced, which provides a driving force for separation.

Though many technologies are available, sophisticated extensive supporting systems are required. Research is in progress to synthesize efficient sorbents and membranes to overcome the challenges associated with these technologies at commercial scale implementation (Bolisetty et al. 2015; Landero et al. 2017).

Oxy-Fuel Combustion

In this process, the combustion processes utilize pure oxygen rather than air. The exhaust gas contains mainly CO_2 and water vapor, from which, water can be easily condensed. Thus, oxy-fuel combustion avoids complex post-combustion separation and also has higher power generation efficiencies. The main disadvantage associated with oxy-fuel combustion is the requirement of pure oxygen as a reactant (Kocs 2017). Most of the studies on oxy-fuel combustion are at laboratory or small scale and these technologies need to be demonstrated at a large scale such as power plants (Ferrari et al. 2017).

Post-Combustion Capture Process

In the post-combustion, CO_2 is separated from fuel gas consisting of high amounts of N_2. Compared with the above two techniques, the addition of this technology with existing power plants is easy (Liang et al. 2016). Membrane technology is also a potential option for post-combustion CO_2 capture due to the inherent advantages: simple in operation, less energy requirement, compact size, stability at high acid concentrations and less water requirement, but the main disadvantage with membrane technology is the permeability which is less than the requirement. And also the large amount of energy is required for vacuum generation.

6.10.6.3 Capture from Atmosphere

CO_2 capture from air (dynamic source), however, is very challenging because of thermodynamic limitations resulting from the extremely dilute CO_2 concentrations, and the energy cost for driving large volumes of air through a capturing process.

6.10.6.4 *Sequestration of CO_2*

Sequestration is the rejection of CO_2, either directly from atmosphere or from anthropogenic sources, and permanent disposing process. A major challenge in mitigating anthropogenic effects on climate change is the reduction of CO_2 emissions to the atmosphere. To this end, no single category of mitigation measures is sufficient, and many of them are mutually dependent and include both reducing greenhouse gas emissions and enhancing greenhouse gas sinks (Bachu 2000; Sillmann et al. 2013).

The global potential and consequences of increased CO_2 consumption are negligible. Some remedial approaches have an incomplete impact without major technological inventions and expenses. If more expensive sequestration approaches are considered, their costs are similar to those for nuclear or renewable energy options, which is not economical. Biomass fixation of CO_2 is a choice for reduction of CO_2 emissions, but due to some cost issues, make it uneconomical for mitigation. Perhaps the oceans signify the largest possible sink, but ocean dumping contains some problems physical and chemical processes which are not understood, cost issues, including sequestration efficiency, technical viability, and environmental influence. Furthermore, ocean disposal also has some political, legal and international limitations for carbon dioxide disposal. Geological sequestration is one of the best process currently, through which we can easily sequester the anthropogenic carbon dioxide, and it is also economical (Bachu and Adams 2003). Criteria used for the CO_2 sequestration are

- Geological
- Hydrocarbon potential and basin maturity
- Hydrothermal and Geothermal
- Economic.

6.10.6.5 *Geological Sinks of CO_2*

Carbon dioxide is heavier and a thermally stable gas than air at normal atmospheric conditions. At 31.18 °C and 7.38 MPa, Carbon dioxide behaves as the gas and in order to obtain the density of Carbon dioxide equal water is dependent upon the temperature and pressure of the CO_2 at 200 to 900 kg/m^3 and it will have obtained the density of water (Hendriks et al. 1993). Solubility of CO_2 is possible by increasing pressure and decreasing the water salinity and temperature because in that cases CO_2 forms solid hydrate which is heavier than water (Holloway and Savage 1993; Koide et al. 1993a, b). In supercritical stage CO_2 is not soluble in water, another way of sequestration is by using the coal beds because carbon dioxide has good affinity toward coal. On the basis of reservoir original pressure and temperature, CO_2 can be deposited either as a compressed gas, liquid or in supercritical phase.

6.10.6.6 Storing in Mine Salt Caverns

This could provide a very long-lasting treatment of CO_2 sequestration in geological media. The technology has already been established and used for underground storage of natural gas, petroleum, and compressed air for industrial and industrial usage. The salt and rock cavern has maximum capacity of storage (Tek 1989; Crossley 1998).

6.10.6.7 Oil Displacement in Reservoirs

Carbon dioxide reduces the interfacial tension and oil viscosity and better solvent for organic compounds. On the base of this property, oil recovery process is used for the sequestration of CO_2 (Herzog et al. 1997). The technology has now been advanced and functionalized on a huge scale in developed sedimentary basins even before climate change became an issue. Enhanced oil recovery (EOR) applications are limited to light crude oil and only miscible displacement is possible. Most of the CO_2 will remain stored in the reservoir, but a significant part eventually breaks through at the producing well, together with the recovered oil and has to be re-circulated back in the system. In turn, the residence time is relatively small (from few months or years). In addition, natural sources can supply CO_2 at a lower cost than from anthropogenic sources (Blunt et al. 1993; Audus 1997).

6.10.6.8 *Use of Methane for Sequestration of CO_2 in Coal Beds*

The affinity of carbon dioxide with methane is twice as compared to coal, as methane is present in huge amount in coal. Two steps are followed, firstly CO_2 adsorbed on coal matrix, in second step methane is produced and it can be used in replacement of coal as much better fuel, for reduction of CO_2 emission (Gunter et al. 1997) (Fig. 6.9).

The majority of coalbed methane resources occurs in India, China, eastern Europe, and Russia, but unfavorable low permeability because of their complex geological location, while USA, Australia, and Canada have large coal reservoirs and are also favorable for sequestration propose (Kelafant et al. 1992; Stevens et al. 1998).

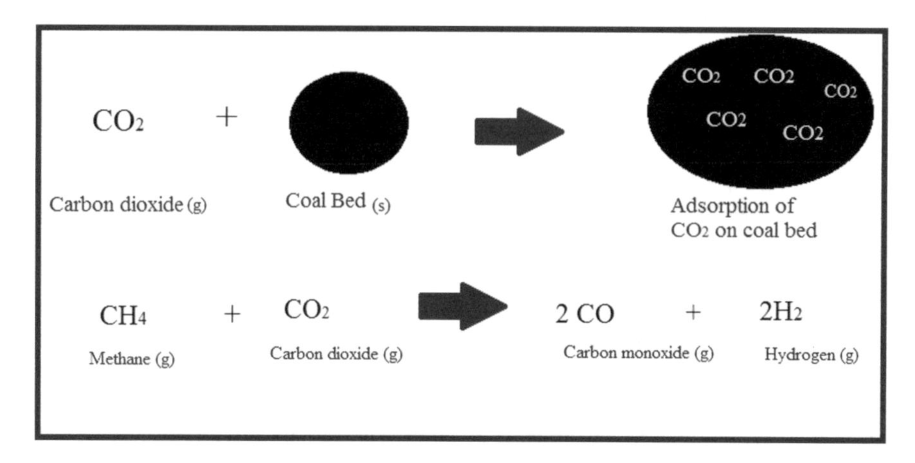

Fig. 6.9 Comparison between Sequestration through coal bed adsorption and methane

6.10.6.9 Storage in Depleted Hydrocarbon Reservoirs

Hydrocarbon reservoirs have been used for CO_2 sequestration, because they have good storing and stopping properties over geological time. The duping mechanism that kept hydrocarbons in the first place should confirm that CO_2 does not reach the surface. The established trap is cheap and economical than other types of CO_2 sequestration as in primary recovery 95% of the CO_2 gas recovered. Oil and gas reservoirs are also used for CO_2 sequestration, but these reservoirs having less potential for CO_2 sequestration. As in the case of ocean disposal, most oil and gas reservoirs are not situated near primary sources of CO_2 production, so new pipelines will be desired to join the CO_2 bases with appropriate sequestration sites. CO_2 sequestration in geological media can be securely happening within national limits, and avoiding international political issues, and usually, there are no environmental problems related with it (John et al. 1984; Bachu and Scherer 1999).

Membrane Technology for Sequestration of CO

Membrane contractors for the absorption of CO_2 are very useful as compared to the traditional gas–liquid contacts. Membrane contacts are beneficial due to specific area, modular unit, independent flow of gas and liquid, and having better internal heat exchange properties. Physical absorption of CO_2 is not a suitable technology due to low partial pressure of CO_2, and chemical absorption is also not considered due to its high energy demand. However, innovative hybrid absorption/stripping membrane contactor (HASMC) for physical solvent carbon capture is proposed, and this process enhanced the absorption of carbon. Furthermore, this technique also fulfills the feasibility of applying physical solvent technology to the treatment of gases with low carbon dioxide partial pressure (Chang et al. 2017) (Fig. 6.10).

The SDGs as a Network of Targets

The main focus of the Open Working Groups originates in the form of 17 goals. Goals have many targets round about 169 targets so, each of these targets are linked with their own goals. According to (Blanc 2015), 2020 attain environmental calm management in case of all the wastes and chemicals. Disposal of chemicals in air and soil, and how their adverse effects can be reduced, which affect the human health and environment. These targets belong to the human health, finance, technology, trading, employment, technology related to water treatment and hygiene. SDG's map as a network of target is shown in Figs. 6.11 and 6.12. The 16 SDG's are represented as large circles and smaller circles which shows the SDG's as network of targets (Figs. 6.13 and 6.14, Tables 6.1, 6.2 and 6.3).

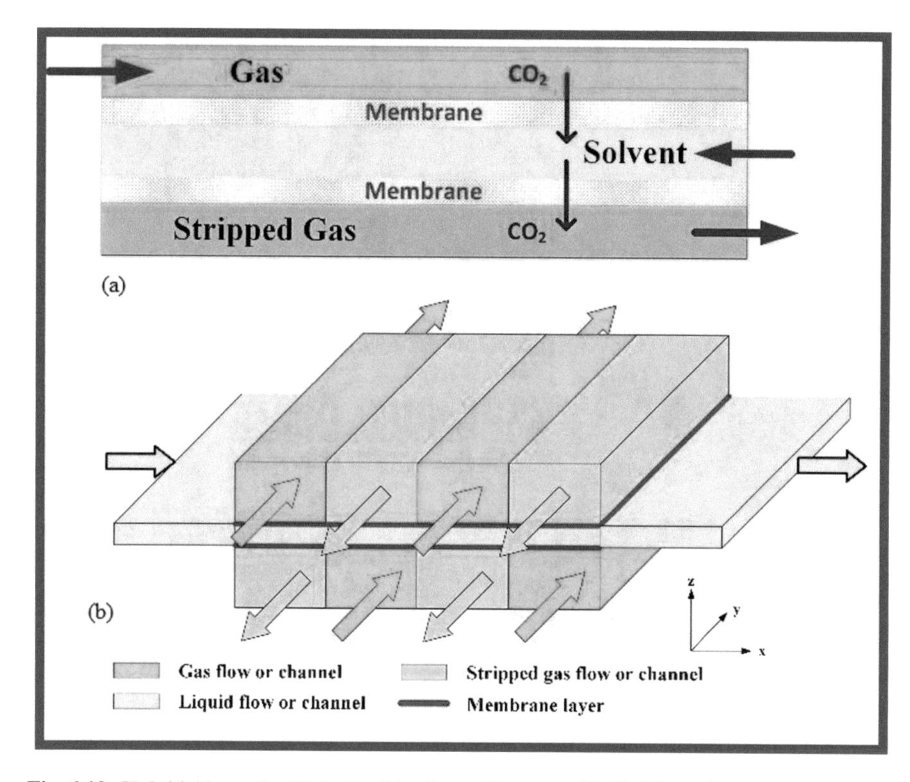

Fig. 6.10 Hybrid Absorption/Stripping Membrane Contactor (HASMC) configurations **a** Parallel-flow; **b** Cross-flow

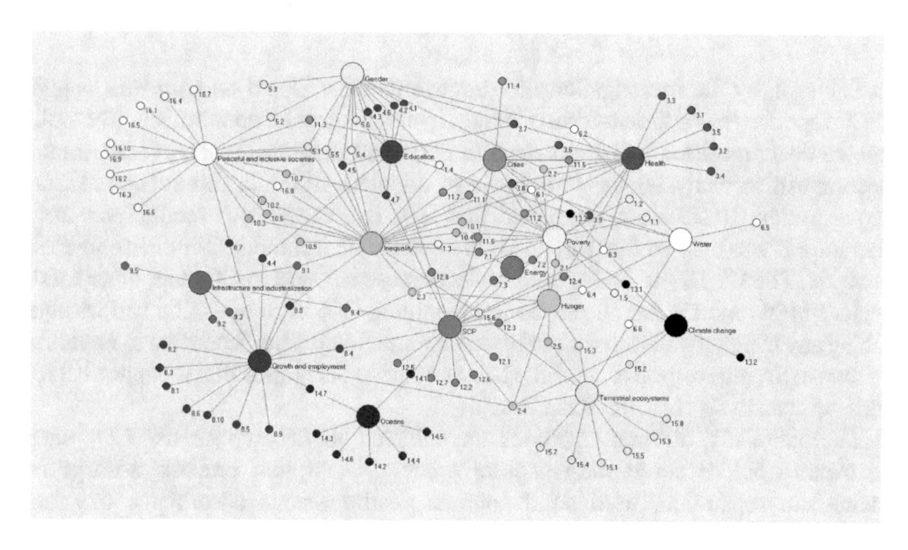

Fig. 6.11 Shows the SDGs as network of targets

Fig. 6.12 Shows the global goals for sustainable development

6.10.7 National and International Report and Process for CO_2 Decomposition

In Pakistan, by the investigation of intergovernmental Board on Climate Change, IPCC reported that different form of energies emitted depend upon source. The well-known decomposition tool introduced in 2001 and the policy such as IDA (index decomposition analysis) was introduced to calculate all the factors related to CO_2 sequestration. Researchers suggested that IDA is an expensive model, and from literature it was cleared that LI and DI are applicable techniques for environmental analysis. The LMDI method is used to decompose changes in CO_2 and other fossil fuels. LMDI report is facing zero issues regarding CO_2 decomposition and without facing any residual problem nationally and internationally. So, according to literature excessive decomposition is helpful whereas additive decomposition is supportive for time-wise analysis (Lin and Raza 2019).

Internationally different methods are adopted to decompose the CO_2 such as capture before combustion capture process, absorption process, adsorption, membrane separation, used solid sorbents, carbonaceous adsorbents, oxy-fuel combustion capture process, etc. (Ahmed and Liu et al. 2020).

Fig. 6.13 Asia and Pacific SDG report

ASIA AND THE PACIFIC
SDG PROGRESS REPORT
2019

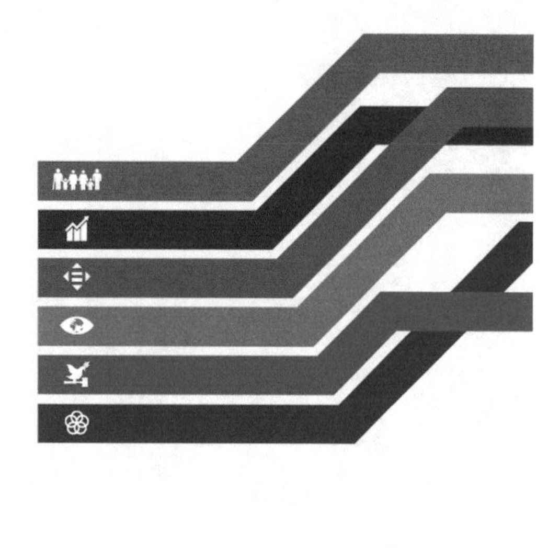

SUSTAINABLE DEVELOPMENT GOALS

6.10.8 Future Prospects

Biotransformation of carbon dioxide in bio electrochemical systems (Fig. 6.15).

Recycling and utilization of CO_2 for the production of chemical and gaseous/liquid energy is the method for reduction of CO_2 in environment. Sequestration of CO_2 is the electrochemical reduction of CO_2 to the fuel. Due to presence of large energy barriers, the surface chemistry of CO_2 reduction is the main task. Therefore, Microbial electro catalyst is the most promising one to reduce the sustainability of carbon level. Bio electrochemical approaches are used to allow the treatment of CO_2. CO_2 can be transformed in terms of process designing and biocatalyst. Furthermore, solar energy is used to convert the CO_2 into valuable fuels and chemical and helpful for the energy saving purpose, and fulfill the energy issues and environmental issues. Development of photocatalytic and photo catalysts will be very essential for the achievement of goals for CO_2 reduction (Bajracharya et al. 2017).

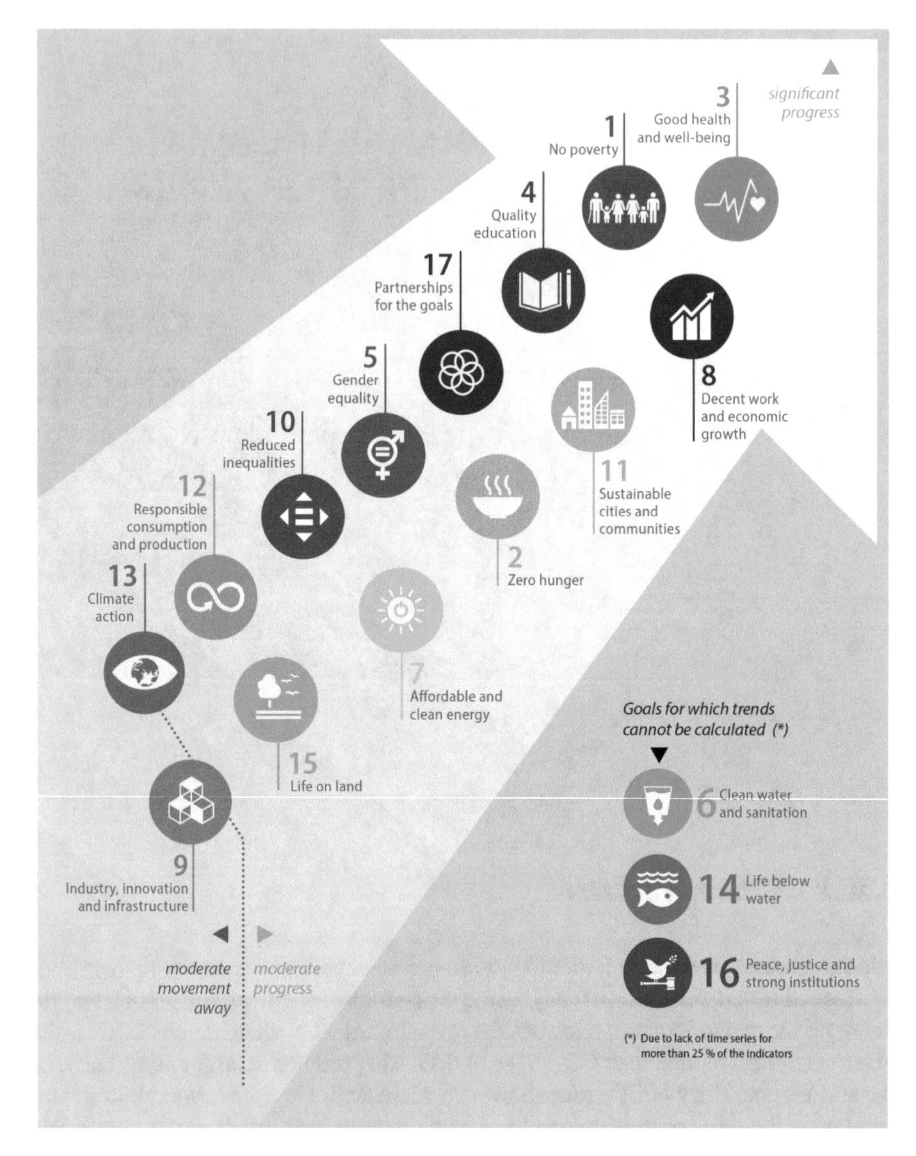

Fig. 6.14 SDG- synopsis-statistics-explained

6.10.9 Conclusion

Distillery waste means the effluent produced due to alcohol distilleries. Alcoholic industries are mostly listed as "Red category." On an average 8–15 L of effluent is generated for every liter of alcohol produced in the alcohol distilleries are extensively growing due to widespread industrial applications of alcohol such as in food,

Table 6.1 Sustainable development goal 7

Sr no	SDG goal (7)	Objective and target of Goal 7	Our objective regarding SDG goal No 7
7	Ensure access to affordable, reliable, sustainable, and modern energy for all	7.1 By 2030, ensure universal access to affordable, reliable, and modern energy services	Sequestration techniques will play a beneficial role to overcome some extent of the environmentalal pollution by burning fossil fuels
		7.2 By 2030, increase substantially the share of renewable energy in the global energy mix	Exposure of waste to the environment, and develop the sequestrations techniques. Clean environment is one of the main goals of SUSTAINABLE DEVELOPMENT GOALS 2015–2030
		7.a By 2030, enhance international cooperation to facilitate access to clean energy research and technology, including renewable energy, energy efficiency and advanced and cleaner fossil-fuel technology, and promote investment in energy infrastructure and clean energy technology	In Pakistan by the investigation of intergovernmental Board on Climate Change, IPCC reported that different form of energys emitted are depending upon source. The well-known decomposition tool introduced in 2001 and the policy such as IDA (index decomposition analysis) was introduced to calculate all the factors related to CO_2 sequestration. Researchers suggested that IDA is an expensive model, from literature it was cleared that LI and DI are applicable techniques for environmental analysis

pharmaceuticals and perfumery. Most of the distilleries exclusively use cane syrup as raw material for fermentation. Sugar syrup is concentrated and then it is diluted with ammonium sulphate, urea, and any other source of nitrogen. Distilleries effluents having high concentrations of Biological oxygen demand (BOD) and Chemical oxygen demand (COD) amounts of spent wash due to the occurrence of huge organic substances. Discharge of CO_2 from distilleries directly affect the environment (atmosphere and oceans ecosystem). The limited concentration of CO_2 is 0.03% which is exceeded due to rapid industrialization. Exceeded level of CO_2 in atmosphere results in climate change which not only affect human health but may result in various changes on earth in the form of different disasters. Among the different sequestration

Table 6.2 Shows the key carbon removal methods

Technology	Description
Afforestation and forest ecosystem restoration	Planting of forests and restoration of ecosystems that result in long-term storage of carbon in above- and below-ground biomass
Bioenergy with carbon capture and storage (BECCS)	Burning biomass for energy generation and capturing and permanently storing the resulting CO_2
Enhancing soil carbon content with biochar	Biomass burning under low-oxygen conditions (pyrolysis) yields charcoal "biochar" which is then added to the soil to enhance soil carbon levels
Enhanced weathering or ocean alkalinisation	Enhancing natural weathering of rocks by extracting, grinding and dispersing carbon-binding minerals on land or by adding alkaline minerals to the ocean to enhance oceanic carbon uptake
Direct air capture and storage	Capturing CO_2 directly from ambient air by a chemical process, followed by permanent storage or use
Ocean fertilization	Fertilizing ocean ecosystems with nutrients to accelerate phytoplankton growth, which partly sinks to the seabed thus moving carbon from the atmosphere to the seabed

Table 6.3 Shows geoengineering methods to solve carbon dioxide discharge issues

Methods	Explanation
Stratospheric Aerosol Injection (SAI)	Injecting reflective aerosol particles or gaseous particle precursors into the lower stratosphere to increase the planetary albedo (reflectivity) and thereby reduce temperatures
Cloud modifications over land or water surfaces	This includes the potential seeding of clouds above ocean surfaces (e.g., with self-steering, autonomous ships), the whitening of clouds above land-surfaces to reflect solar radiation away from earth and the thinning of cirrus clouds to allow more heat to escape
Surface albedo modifications	Making various surfaces such as urban areas, roads, agricultural land, grasslands, deserts, polar ice caps or oceans brighter to prevent solar radiation from heating up the areas covered

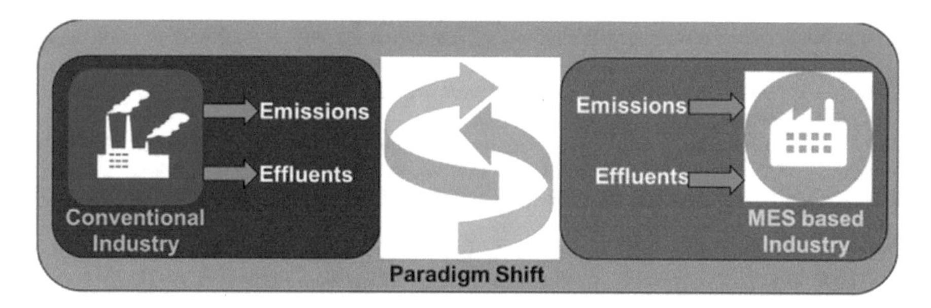

Fig. 6.15 Graphical representation of conventional and modified industry for CO treatment

processes used for the CO_2 treatment such as geological processing and sequestration through membrane technology are more economical.

References

Ahmed R, Liu G et al (2020) Recent advances in carbon-based renewable adsorbent for selective carbon dioxide capture and separation-a review. J Cleaner Prod 242:118409

Audus H (1997) Greenhouse gas mitigation technology: an overview of the CO_2 capture and sequestration studies and further activities of the IEA Greenhouse Gas R&D Programme. Energy

Bachu S (2000) Sequestration of CO_2 in geological media: criteria and approach for site selection in response to climate change. Energy Convers Manage 41(9):953–970

Bachu S, Scherer GW (1999) Storage capacity of CO_2 in geological media in sedimentary basins, with application to the Alberta basin. In: Proceedings of the 4th international conference on Greenhouse Gas Control Technologies (GHGT-4), Interlaken, Switzerland. Elseveir

Bachu S, Adams JJ (2003) Sequestration of CO_2 in geological media in response to climate change: capacity of deep saline aquifers to sequester CO_2 in solution. Energy Convers Manage 44(20):3151–3175

Bajracharya S, Srikanth S, Mohanakrishna G, Zacharia R, Strik DP, Pant D (2017) Biotransformation of carbon dioxide in bioelectrochemical systems: State of the art and future prospects. J Power Sources

Bern C1, Sniezek J, Mathbor GM, Siddiqi MS, Ronsmans C, Chowdhury AM, Choudhury AE, Islam K, Bennish M, Noji E, et al (1993) Risk factors for mortality in the Bangladesh cyclone of 1991. Bull World Health Organ

Blanc DL (2015) Towards integration at last? The sustainable development goals as a network of targets. Sustainable Development

Blunt M, Fayers F, Orr FM (1993) Carbon dioxide in enhanced oil recovery. Energy Convers Mgmt

Bolisetty S, Pathi S, Inkollu S, Kondapuram VR (2015) Development trends in porous adsorbents for carbon capture. Environ Sci Technol

Brown M (2009) The response of northern hemisphere snow cover to a changing climate

Chams N, García-Blandón N (2019) On the importance of sustainable human resource management for the adoption of sustainable development goals. Resour, Conserv Recycl

Chang H, Gan H-Y, Pan R-H, Ho C-D (2017) CFD study of hybrid membrane contactors for absorption and stripping of carbon dioxide. Energy Procedia

Corinne Le Quéré MRR, Canadell JG, Marland G (2009) Trends in the sources and sinks of carbon dioxide. Nat Geosci

Critoph RE (1988) Performance limitations of adsorption cycles for solar cooling. Sol Energy 41(1):21–31

Crossley NG (1998) Conversion of LPG salt caverns to natural gas storage "A Transgas experience" J Can Pet Technol

Dale VH (1997) The relationship between land-use change and climate change. Ecological Processes

Edward B (1997) Climate process and change. Cambridge University Press, Cambridge, UK, p 209

EPRI (1995) Potential effects of climate change on electric utilities. Electric Power Research Institute

Feeley TJ, Skone TJ, Stiegel GJ, McNemar A, Nemeth M, Schimmoller B, Murphy JT, Manfredo L (2008) A critical resource in the thermoelectric power industry. Energy 33:31–11

Femke Vos JR, Below R, Guha-Sapir D (2009) Annual disaster statistical review 2009

Ferrari N, Mancuso L, Davison J, Chiesa P, Martelli E, Romano MC (2017) Oxyturbine for power plant with CO_2 capture. Energy Procedia

Fritz SC (1996) Paleolimnological records of climate change in North America. Limnol Oceanogr

Fujii H, Iwata K, et al (2017) How do urban characteristics affect climate change mitigation policies? J Cleaner Prod 168(Supplement C):271–278

Gleick PH (1994) Annual review of energy and the environment. Water and Energy 19:267–299

Gunter WD, Gentzis T, Rottenfusser BA, Richardson RJH (1997) Deep coalbed methane in Alberta, Canada: a fuel resource with the potential of zero greenhouse emissions. Energy Convers Mgmt

Hansen J, Ruedy R, Sato M, Lo K (2010) Global surface temperature change. Rev Geophys

Hansen J, Johnson D, Lacis A, Lebedeff S, Lee P, Rind D, Russell G (2017) Climate impact of increasing atmospheric carbon dioxide. American Association for the Advancement of Science

Hare B, Schaeffer M, Rocha M (2011) Science aspects of the 2 °C and 1.5 °C global goals in the Cancun Agreements, LDC paper series. Climate Analytics, Telegrafenberg, Germany

Hay M, Thomas DW, Craighead JL, Economides C, Rosenthal J (2006) Clinical development success rates for investigational drugs. Nat Biotechnol

Hendriks CA, Bakasis I, Karampasis MM (1993) Underground storage of carbon dioxide. Energy Convers Mgmt

Herzog HJ, Drake E, Adams EE (1997) CO_2 capture, reuse, and storage technologies for mitigating global climate change. Final Report, DOE No. DE-AF22–96PC01257, Massachussets Institute of Technology, Cambridge, MA, p 66

Holloway S, Savage D (1993) The potential for aquifer disposal of carbon dioxide in the UK. Energy Convers Mgmt

Hotez PJ,· Brindley P, Bethony JM, King CH, Pearce EJ, Jacobson J (2008) Helminth infections: the great neglected tropical diseases. J Clin Invest

Hurd B, Harrod M (2001) Water resources: economic analysis. In: Mendelsohn R (ed) Global warming and the American economy. Edward Elgar Publishing Ltd., Cheltenham, pp 106–131

IEA (2017) Renewable information: overview. International Energy Agency

IEA, GEO (2013) Understanding the electric vehicle landscape to 2020

Intergovernmental Panel on Climate Change (1991) Climate change: the IPCC response strategies. Island Press, Washington, DC, p 272

IPCC (2007) Climate change 2007: impacts, adaptation and vulnerability. Cambridge University Press, Cambridge

Jacoby HD (1990) Climate change and U.S. water resources. In: Waggoner PE (ed) Wiley, New York. Water Quality

Jepma CJ, Munasinghe M (1998) Climate change policy. Cambridge University Press, New York, p 331

John BS, Bally A, Klemme HD (1984) Sedimentary provinces of the world Ð hydrocarbon productive and nonproductive. The American Association of Petroleum Geologists

Justic D, Rabalais NN, Turner RE (1996) Effects of climate change on hypoxia in coastal waters: a doubled CO_2 scenario for the Northern Gulf of Mexico. Limnol Oceanogr

Kampa M, Castanas E (2008) Human health effects of air pollution. Environ Pollut

Kanae S, Kundzewicz ZW, Hirabayashi Y (2010) River floods in the changing climate—observations and projections. Water Resour Manage
Kaplan YA (2015) Overview of wind energy in the world and assessment of current wind energy policies in Turkey. Renew Sustain Energy
Karakosta C, Pappas C, Marinakis V, Psarras J (2013) Renewable energy and nuclear power towards sustainable development: characteristics and prospects. Renew Sustain Energy
Kelafant JR, Stevens S, Boyer II CM (1992) Coalbed gas-2: vast resource potential exists in many countries. Oil Gas J
Kiron D, Kruschwitz N, Haanaes K, von Streng Velken (2012) Sustainability nears a tipping point. MIT Sloan Manage Rev 53(2):69–74
Kirshen P, Ruth M, Anderson W (2008) Interdependencies of urban climate change impacts and adaptation strategies: a case study of Metropolitan Boston Clim Change
Koch H, Vögele S (2009) Dynamic modelling of water demand, water availability and adaptation strategies for power plants to global change. Ecol Econ 68(7):2031–2039
Kocs EA (2017) The global carbon nation: status of CO_2 capture, storage and utilization. In: EPJ web of conferences
Koide HG, Tazaki Y, Noguchi Y, Iijima M, Ito K, Shindo Y (1993) Carbon dioxide injection into useless aquifers and recovery of natural gas dissolved in fossil water. Energy Convers Mgmt
Koide HG, Tazaki Y, Noguchi Y, Iijima M, Ito K, Shindo Y (1993) Underground storage of carbon dioxide in depleted natural gas reservoirs and useless aquifers. Eng Geol
Landero AC, Gutiérrez AJ, Gani R (2017) Development of an intensified reactive distillation process for the synthesis of dioxolane products. Comput-Aided Chem Eng Geol
Liang Z, Fu K, Idem R, Tontiwachwuthikul P (2016) Review on current advances, future challenges and consideration issues for post-combustion CO_2 capture using amine-based absorbents. Chin J Chem Eng
Lin B, Raza MY (2019) Analysis of energy related CO_2 emissions in Pakistan. J Cleaner Prod 219:981–993
Luber G, Lawrimore J (2015) Global climate change and human health
McKnight DM, Brakke DF, Muiholland PJ (1996) Freshwater ecosystems and climate change. Limnol Oceanogr
Mohana S, Acharya BK et al (2009) Distillery spent wash: treatment technologies and potential applications. J Hazard Mater 163(1):12–25
Mueller DK, Hamilton PA, Helsel DR, Hitt KJ, Ruddy BC (1995) Nutrients in groundwater and surface water in the united states—an analysis of data through 1992. U.S. Geological Survey Water Resources Investigation Report
Nielson EG, Lee LK (1987) The magnitude and costs of groundwater contamination from agricultural chemicals. Agricultural Economic Report No. 576, USDA Resources and Technology Division, Economic Research Service, Washington
Pascual M, Ahumada JA, Chaves LF, Rodó X, Boum M (2006) Malaria resurgence in the East African highlands: temperature trends revisited. PNAS
Postel SL, Daily GC, Ehrlich PR (1996) Human appropriation of renewable freshwater. Science
Rahman FA, Aziz MMA, Saidur R, Bakar WAWA, Hainin MR, Putrajaya R, Hasan NA (2017) Pollution to solution: capture and sequestration of carbon dioxide (CO_2) and its utilization as a renewable energy source for a sustainable future. Renew Sustain Energy Rev
Ribeiro SK, Rosa LP (2017) Avoiding emissions of carbon dioxide through the use of fuels derived from sugar cane. Springer on behalf of Royal Swedish Academy of Sciences 27
Richardson K, Steffen W, Abernathy DL (2011) Climate change: global risks, challenges and decisions. Cambridge, p 247
Roh K, Frauzem R, Ganib R, Lee JH (2016) Process systems engineering issues and applications towards reducing carbon dioxide emissions through conversion technologies: review. Chem Eng Res Des
Rubin ES (2005) IPCC special report on carbondioxide capture and storage

Sabir A, Islam A, Shafiq M, Shafeeq A, Butt MTZ, Ahmad NM, Sanaullah K, Jamil T (2014) Novel polymer matrix composite membrane doped with fumed silica particles for reverse osmosis desalination

Sabir A, Shafiq M, Islam A, Sarwar A, Dilshad MR, Shafeeq A, Butt MTZ, Jamil T (2015) Fabrication of tethered carbon nanotubes in cellulose acetate/polyethylene glycol-400 composite membranes for reverse osmosis. Carbohydr Polym

Sabir A, Shafiq M, Islam A, Jabeen F, Shafeeq A, Ahmad A, Butt MTZ, Jacob KI, Jamil T (2016a) Conjugation of silica nanoparticles with cellulose acetate/polyethylene glycol 300 membrane for reverse osmosis using $MgSO_4$ solution

Sabir A, Shafiq M, Islam A, Jabeen F, Shafeeq A, Ahmad A, Butt MTZ, Jacob KI, Jamil T (2016b) Conjugation of silica nanoparticles with cellulose acetate/polyethylene glycol 300 membrane for reverse osmosis using $MgSO_4$ solution. Carbohydr Polym

Sabir A, Falath W, Jacob KI, Shafiq M, Gull N, Islam A, Munawar MA, Zia S, Khan SM, Shafeeq A, Butt MTZ, Jamil T (2016c) Integrally skinned nano-cellular crosslinked asymmetric thin films infused with PEO-PPO-PEO block copolymer/ZnO-NPs for desalination using sea salt. Mater Chem Phys

Sabir A, Falath W, Jacob KI, Shafiq M, Munawar MA, Islam A, Gull N, Butt MTZ, Sanaullah K, Jamil T (2016d) Hyperbranched polyethyleneimine induced polycationic membranes for improved fouling resistance and high RO performance

Sawyer JS (1972) Man- made carbondioxide and green house effect. Nature

Sillmann J, Kharin VV, Zhang X, Zwiers FW, Brona D (2013) Climate extremes indices in the CMIP5 multimodel ensemble: Part 1. Model evaluation in the present climate. J Geophys Res

Stevens SH, Spector D, Riemer P (1998) CO_2 sequestration in deep coal seams: pilot results and worldwide potential. In: Proceedings of the 4th international conference on Greenhouse Gas Control Technologies (GHGT-4), Interlaken, Switzerland. Elseveir

Stocker TF, Qin D, Plattner G-K, Tignor M, Allen SK, Boschung J, Nauels A, et al (2013) Climate change 2013: the physical science basis .contribution of working group I to the fifth assesment report of the intergovernmental panel on climate change. Cambridge University Press, Cambridge

Tek MR (1989) Underground storage of natural gas: theory and practice. In: NATO ASI series E, applied sciences

Viswanathan L, Kumar S (1991) Production of biomass, carbon dioxide,volatile acids, and their interrelationship with decrease in chemical oxygen demand, during distillery waste treatment by bacterial strains. Enzyme Microb Technol

Watson RT, Zinyowera MC, Moss RH (1996) Impacts, adaptations and mitigation of climate change: scientific-technical analyses. Cambridge University Press, Cambridge, Massachusetts

Werner PC, Gerstengarbe FW (1997) Proposal for the development of climate scenarios. Clim Res

Wurbs RA (2005) Modeling river/reservoir system management, water allocation and supply reliability. J Hydrol 300:100–113

Zebisch M, Grothmann T, Schröter D, Hasse C, Fritsch U, Cramer W (2005) Climate change in Germany: vulnerability and adaptation of climate-sensitive sectors. Umweltbundesamt Clim Change

Chapter 7
Materialization of CO_2 from Distilleries in Algae-Based Biofuel and Biomass

Uqba Mehmood and Muhammad Faisal

Abstract The continuous rise in CO_2 emission is a big reason behind global warming, with the passage of time, it is getting worse. Research on algal bioenergy has opened many ways for energy production as well as reducing CO_2 in the environment by capturing it with its unique systems. In 1942 Harder and von Witsch unveiled the idea of microalgae growth to obtain lipids either in the form of food or fuel. Algae can be used to get or might be converted to different types of fuels. To achieve this objective, variable techniques can be applied, depending upon the phase and kind of cell being used. This chapter will add to the comparison of reliability of various procedures, for treatment of distillery wastewater prior to utilization as a nutrient source for algae. *Chlorella vulgaris* SR/2, a strain has been used for the study of materialization of CO_2 from distilleries. It was concluded from the study that the distillery water priory treated with biological techniques have more ability for the better growth of algae. The environment is accommodated with variable amounts of CO_2, luckily microorganism who can synthesize their own food by using carbon such as cyanobacteria and algae are able to survive in both high and low levels of carbon dioxide. Excellency of the pathway is more enhanced with carboxysome or the pyrenoid like structures which carry Rubisco. Oil is extracted and the other remnants of algae are pelletized after drying. These pelletized remnants are used in industry for power generation after being burnt.

Keywords Sustainable development goal 7 · Biomass · Algae · Cultivation · Extraction

- Agenda 2030 by United Nations focuses on enhancing the efforts to reduce emission of global greenhouse gases and address adverse effects of climate change.
- In research by Andersson et al. (2014) municipal water is used to cultivate algae for the production of biofuel.

U. Mehmood (✉)
Department of Allied Health Sciences, The Superior College, Raiwind Road, Lahore, Pakistan
e-mail: uqba.mehmood@superior.edu.pk

M. Faisal
Department of Microbiology and Molecular Genetics, Punjab University, Lahore, Pakistan

M. Arshad (ed.), *Sustainable Ethanol and Climate Change*,
https://doi.org/10.1007/978-3-030-59280-6_7

Table 7.1 Data on energy consumption

Facts	
Developing and underdeveloped countries lack electricity in rural areas	1 in 7
Greenhouse gas production by energy which interferes with climate change	60%
Efficient ways of energy production standards can reduce the consumption of electricity in building and industry	14%
Low-standard fuels, which affect health, are used for cooking by people	3 billion
Use of renewable sources to generate power in 2015	20%
Employment in the sector of renewable energy in 2017	10.3 million

- Several methods for treatment of distillery effluent before carbon removal are applied.
- Globally, a consensus was observed on biofuel production to maintain sustainability.
- Before using marine algae biomass for large-scale biofuel production, several technological issues need to be addressed.
- Large corporations are investing billions of dollars, in both developed and under develop nations.

7.1 Introduction

Demands to explore new technology and strengthen the infrastructure to fulfill energy needs of all without interfering climate are getting high (Table 7.1). Beginning of twenty-first century, human population consuming electricity becomes 78%, while remaining who are not consuming electricity dropped to one billion. With the rise in growth of population, reciprocally raise the demand for energy, which ultimately enhance the worst changes in the environment by relying on fossil fuels economically. The alternative ways for increasing energy productivity are solar, wind, and thermal power, which ensures to get the target of SDG-7 by 2030.

7.2 Sustainable Development Goal 7 (SDG-7) and Renewable Energy

Agenda 2030 by United Nations for sustainable development is comprised of mutual association at international level; it focuses on enhancing the efforts to reduce the emission of global greenhouse gases and address to cope with climate changes which have adverse effects. The main objective is to be saved from the interference of greenhouse gas with the climate by maintaining its level at a particular point (weblink 1; weblink 2). There were many yearly sessions of Commission on Sustainable Development (CSD) on this issue. The major sessions were in 1996 on the protection of atmosphere, while in 2006 and 2007, sessions were on air and atmospheric pollution. A conference on Climate Change (COP24) in Poland was held in December 2018. The Intergovernmental Panel on Climate Change (IPCC) presented 30 sessions regarding 1.5 °C, emphasizing on threats of climate change. The report showed that global warming exceeds 1.0 °C due to adverse activities of humans. The threat prevails that it can be reached to 1.5 °C in 2030 or 2050 (weblink 3).

Now, the main focus was to work on renewable energy sources which are natural resources that cannot be extinct. Renewable energy can replace other traditional energy which is fossil fuel and rather it is more environment-friendly.

Seven types of renewable sources are defined as following:

Solar energy converts sunlight to heat or electricity. Solar panels use solar cells to capture light and convert it into electricity through photovoltaic system. Large space and maintenance of panels required.

Wind farms convert wind energy to electricity via turbines. Wind power is generated through the variation in temperature in the atmosphere, so we can consider it a form of solar system. There are various systems to generate electricity from wind power. Although it is environment-friendly, they must be built in barren areas, away from cities.

Hydroelectric is associated with dams. A waterfall is managed to flow through turbines to generate electricity. A huge landscape setup is required to achieve results.

Geothermal heat is generated after a long duration decay of radioactive process released in the form of steam, volcanic eruption, etc., through operating turbines. It is not only costly but also dangerous to deal with.

Ocean can produce thermal (warm water surface temperature) and mechanical (flow of tides) energy through various systems. The output is conditional and, in normal state, very less.

Hydrogen reaction to form water generates heat which can be optimized for fuel and electricity. It is unable to prevent pollution.

Biomass is the most effective source to generate energy called bioenergy. It involves living plants and organisms; different methods are applied to get energy from biomass. Those are maybe burning and decomposition (disintegration under pressure). It is all about maintaining a balance between consumption and production of carbon dioxide which ultimately balances the atmosphere. The energy stored during this whole process is from the sun in the form of chemical energy (through photosynthesis) and later on released in the form of heat. Liquid biofuels (biodiesel and ethanol) and biogas are the end products burned directly to get different forms of energy.

Importance of energy production from biomass can be analyzed considering the example of the United States, where, overall, 5% of basic energy depends on biomass fuels in 2017. Among that, 47% was from biofuel like ethanol, municipal waste 10%, while 44% depended on wood and its derivatives. It is safe and environment-friendly; so many researches are focusing on the use of biomass for fuel production (Fig. 7.3).

7.3 Trends to Invest in Renewable Energy Sources to Achieve SDG-7

Currently, 3 billion population is lacking clean and affordable energy and are using alternatives which are exposing them to highly polluted levels of hazards. Less than 1 billion are out of reach to electricity, 50% of them are in Africa and sub-Sahara. The work on renewable sources has declined the per GDP unit of energy use. There is a significant potential in biomass to meet the needs of energy supply with high-energy demands of populous nations like China, Brazil, and India (Table 7.2).

However, there is no sufficient energy available yet to fulfill all needs. More research is required to introduce new options for trial. The integration of renewable energy, in other applications such as daily use, infrastructure, gray material, mobility, and industry, is required. Transformation of energy systems in the whole world must be increased through investment on priority in regulating and creating new business models.

Bioenergy is one of the main pillars of renewable energy sources. It is a solution to most of the challenges which are created continuously. Those challenges are lack of professional growth, safety risk, changes in climate, food scarcity, and energy unavailability. To achieve this, goal 7 is important as it is interlinking opportunity to answer all the above challenges. The access to energy through bioenergy will enhance energy efficiency while improving economy and in return job opportunities, in parallel addressing changes in climate.

All member countries of the United Nations Organization show their consensus on the agenda of protection and development of people on this planet till 2030. The agenda consists of 17 goals among which goal 7 is all about clean and affordable energy. To achieve this target, the whole focus is required to invest in renewable energy sources, for researching and enhancing efficiency to expand access to energy without disturbing nature. Atmosphere conservation is a broad and diverse endeavor comprising multiple areas of financial activity. In this reference, the United Nations ensures to mobilize jointly $100 billion annually by 2020. The United Nations secures applications and utilization of the Green Climate Fund through its capitalization in all possible ways (weblink 1; weblink 2).

Table 7.2 World production capacity and production (Ethanol Fuel and Biomass-based diesel)

	2010	2011	2012	2013	2014	2015
World						
Production (Mb/d)	1842.976	1903.612	1972.565	2119.533	2260.328	2285.22
Fuel ethanol (Mb/d)	1478.596	1454.582	1490.045	1582.293	1662.618	1737.29
Biomass-based diesel (Mb/d)	364.38	449.03	482.52	537.24	597.71	547.93

U.S. Energy Information Administration | Monthly Biodiesel Production Report

Brazil is leading in the production of liquid biofuels. It owns the largest vehicles and fleet which are consuming bioethanol. Bioethanol is produced by the fermentation of carbohydrates. The carbohydrates from crops, e.g., sweet sorghum, sugarcane, and corn, are used. Then Pakistan, India, United Kingdom and Argentina are making their place.

Recently, many magnificent funding bodies switch their funds from fossil fuels to renewable energy sources, especially bioenergy. The change in investment trend is remarkable and highlighted on media. Many investors take it as their "moral responsibility to preserve a healthy planet" such as Valerie Rockefeller Wayne (Trustee of Rockefeller fund) whom the Washington Post quoted while admitting it.

Google with investors like GE, ConocoPhillips, BP, and Google Ventures is working on "Cool Planet Project." They are using pyrolyzer (thermomechanical fraction system) to get hydrocarbon chains (in the form of gases) from wood waste and energy crops. These gases will raise the energy level of jet fuels, diesel, and gasoline.

Pension funds of £160 m were also invested in a biomass power plant known as Brigg Renewable Energy Plant (weblink 21).

7.4 Role of Algae as a Renewable Energy Source

One of the reliable tools to control climate change is Carbon Dioxide Removal (CDR). It is reported that the phenomena that limit global warming to 1.5 °C are depending on the use of CDR at any extent. The impact of CDR directly or indirectly enhances the quality of land, food and water security, ecosystems, and biodiversity (weblink 4). Algae have been used to remove carbon dioxide from different aquatic systems. The concept to use algae for different purposes is not recent. It is used by more than 50 years ago. The use of methane gas was a blessing during the crisis of energy in 1970. Methane gas production from algae was initiated in 1950, but its role was significantly appreciated later on. The criteria are getting changed in time. Needs regarding fuel are increasing globally, most of the countries are increasing import of fuel, and high-alert securities are required for all these. These issues are forced to focus on establishing biofuel systems particularly algae-based biofuels. Algae are studied in-depth, to be used in advance technology, to extract biofuels (Demirbas 2010; Wijffels and Barbosa 2010; Kilian et al. 2011). In recent meetings regarding targets for global change in climate, a consensus was observed on biofuel production to maintain sustainability (weblink 5; weblink 6). Primitive or conventional ways to use fuel are the cause of many issues related to environment. Main shortcoming of conventional fuels, either starch-based or extracted from plants, is the engagement of land and the treatment of CO_2-containing wastewater and quantity of biofuel production, which is compared during microalgal-based research (Paswan and Nikhil 2014; weblink 7; weblink 8). The Department of Energy in the USA allocates a lot of time and money to establish Aquatic Species Program (ASP). The main purpose was to be enabled to produce oil from algae. It took approximately 20 years with an amount of 25 million dollars.

With benefits, some hazards related to the use of algae were also reported by NREL; it is a scientific community working on algae-based fuels on a smaller level. They inhibited to proceed with this technology, quoting "The study of microalgae represents an area of high risk and high gains." Despite the risks are mentioned on economical basis, still, several companies in the United States are exploring ways to use algae to reduce greenhouse gas emissions from power plants and to generate biodiesel. These companies include GreenFuel, Solazyme, and LiveFuels (weblink 9).

Solazyme is a San Francisco-based company. It has modified the cultivation techniques to grow algae. It cultures genetically modified algae in dark providing sugar as the nutritive value in large containers. This modification helps to get more oil and growth of algae in high concentration, after boosting the metabolic processes which are not active otherwise.

GreenFuel Technologies of Cambridge is one more stakeholder in the algae-based fuels market. GreenFuel has conducted a trial run at the Massachusetts Institute of Technology's (MIT) Cogeneration power plant which showed a reduction of CO_2 emission, nearly 82% with 85% decrease in nitrous oxide in algae system (weblink 10).

Other than the United States, Aquaflow Bionomic Corporation (ABC) of New Zealand has adopted a different approach to algae-based biofuels. They are cultivating algae in effluent system settling ponds and other water bodies rich in nutrients. Aquaflow Bionomic pursues to develop cost-effective methods simultaneously cleaning environment with the production of biofuels (weblink 11).

Nowadays, the Iranian Government has focused a lot of attention exploring the ways to use renewable energy, especially in the field of biofuel production. A large area unable to cultivate for food is good to grow algae. Lake Orumieh in Iran's West Azarbaijan province, Maharlu lake in Iran's Fars province, Qom lake in Iran's Qom province is a salt lake to be used to cultivate algae. New species of algae are grown to extract biofuels (Najafi, et al. 2011).

An interesting report is received from Andersson et al. 2014. In this research work, municipal water is used to cultivate algae for the production of biofuel. The study was opted in Gothenburg, Sweden. Two setups were designed on the basis of biorefinery concept. In setup A, combined biodiesel and biogas production, while in setup B only biogas production. In this study, not only production rates and impact on global CO_2 emissions mitigation were recorded but also the efficiency of biofuels from algae and other productive ways was compared. It focused on the collaboration among all reactors such as algae cultivation, biofuel production processes, a wastewater treatment plant, and an industrial cluster. The rate of CO_2 emissions reduced in collaboration as compared to alone operation. It also highlighted that biofuel production from algae is more efficient than other ways to get biofuel (Andersson et al. 2014).

The production of biodiesel from algae is an advanced level of biofuel production which is preferable due to production of biomass in a very short period and the land used can be non-arable (Dragone et al. 2010). The 10% production of biofuel from microalgae is enough to cover the expenditures of nutrients required for the growth of algae (Delrue et al. 2016). This cost is manageable with the utilization of nutrients present in wastewater, either from the distillery or other effluents. The major nutrients which play a vital role in microalgae cultivation are nitrogen and phosphorous are rich in wastewater. The advantage of microalgae is not only the utilization of nutritious compounds but also the removal of organic pollutants (hormones and pharmaceutical wastes) and heavy metals from wastewater (Delrue et al. 2016; Subashchandrabose et al. 2013).

The potential to act as feedstock for biofuel, specifically biodiesel, in comparison to other conventional sources, is high in microalgae. Up to 80% of lipids can be extracted from the dry weight of microalgae (Chisti 2007). One drawback to achieve high quantity of lipids is exerting stress through variable environmental conditions. The change in environment, either physical or chemical, ultimately slows down the growth rate by reducing cell division. All these factors ultimately cause low production of biomass. This vice versa relation unable one's to get high quantity lipid and large amount of biomass simultaneously (Hu et al. 2008). The best way is to divide the procedure into two phases: in the first phase, let the microalgae grow extensively, while in second phase, stress is applied for lipid. It is the only possible way to achieve both biomass and lipid. The selection of engineered microalgae to obtain lipid is another option, without affecting the yield of biomass (Trentacoste et al. 2013).

After a wide research on the use of microalgae to get lipid, it was concluded that all strains cannot be used for the production of biodiesel. Biodiesel obtained from microalgae source is unsaturated at a high level which characterizes it in low oxidative stability in comparison with biodiesel accumulated from other sources. This drawback can be overcome by accurate mixing of it with other chemical stabilizers or fossil diesel (Kumar and Sharma 2016; Ribeiro et al. 2007; weblink 12). Temperature has various effects on diesel properties, which eventually affect the performance of fuel (Bhale et al. 2009). Generally, there are many important properties of diesel which make it suitable for devices according to use including the energy potential, ratio of water and acid, etc. These qualities of diesel characterize the corrosiveness of oil. Viscosity is one of the characteristics which influences the performance of engine (weblink 12).

The first report on integrated wastewater treatment and biodiesel production in a native cold climate with alga is recently reported by Jämsä et al. (2017). The algae selected for this study showed high-level removal nutrients, synthesis of lipids and biomass in comparison to other strains (Lynch et al. 2015). It was Finnish isolate UHCC0027, affiliated to family Scenedesmaceae. Growth analysis of algae and level of removing nitrogen and phosphorous were compared at variable Chemical Oxygen Demand (COD) and nitrogen:phosphorous ratios. The amount and chemical composition of fatty acids were evaluated later on. The fall in temperature exhibited a very low effect on COD removal, while growth depends on temperature which is ultimately a requirement for the removal of nitrogen and phosphorous (main role is of N:P ratios).The hydraulic retention time (HRT) of 12 days is required for culture regulation to achieve nitrogen and phosphorous removal. The quality of biodiesel was affected by low temperature. The extracted fatty acids were polyunsaturated (Jämsä et al. 2017).

Advance biotechnology research, such as the synthesis of genetically modified algae (Kwaka et al. 2017) which is used to obtain oil, is able to utilize solar energy more efficiently; ultimately, it provides innovative ways to explore. Achieving high benefits from commercialization of biofuel from algae needs upgradation in all phases involved in biofuel synthesis, including algal growth improvement, algae culturing, collecting, oil extraction, and conversion to variable products (Fon Sing et al. 2011; Mata et al. 2010). Although almost whole work regarding algae, from production to commercialization, is carried on in the United States. Large corporations are investing billions of dollars, along with the US Government. However, now algae-based biofuel production is being underresearch in Europe, Asia, and elsewhere, in both developed and underdevelop nations.

7.5 Pretreatment Methods of Distillery Wastewater

There are several different methods for the treatment of distillery effluent. They are as follows.

7.5.1 *Physiochemical Treatment Methods*

- Coagulation: Reduction of repulsive forces by the addition of coagulant.
- Flocculation: Physical process by which particle contact and agglomeration occur.
- Ion Exchange: To separate ionized molecules (organic as well as inorganic) from aqueous solution as well as contaminants in organic streams.
- Hydrodynamic Cavitation Technology: Cavitation is a process in which cavities or bubbles are formed and then collapsed ultimately creating oxidizing agents in wastewater.
- Membrane Technology: Distillery effluents from industry is of acidic nature nearly of pH range 3–5. To neutralize it, sodium hydroxide is added which results in a lot of suspended particles. To remove those particles, thin-pored cloth or membrane is used, so with the help of fine filtration all suspended particles are removed.

7.5.2 *Biological Treatment Methods*

- Anaerobic Treatment.
- Aerobic Treatment.
- Enzymatic Treatment.
- Bacterial Role.

Biological treatment methods are considered as the efficient mode of pretreatment in case of industrial wastewaters, which are highly polluted. Both anaerobic and aerobic techniques are already applied to improve the state of effluents from industries.

7.6 Anaerobic Treatment

The most advisable primary treatment is anaerobic digestion of distillery wastewaters because aerobically treated methods require high amount of nutrients, with this in the presence of high levels of organic compounds, cause difficulties during operation. Aerobic treatment can only be helpful in the case of primary treatment of stillage; otherwise, it can be an expensive method. Most of the anaerobic pretreatment applications, which have been proved significant in the field of wastewaters in municipal and other industry branches (which were rich in organic compounds), were also applied for wastewater from ethanol plants; the results were highly reliable for pollutant decay (Rani et al. 2013). Tomczak-Wandzel et al. (weblink 13) reported that UASB (upflow anaerobic sludge blanket reactor), a treatment by anaerobic technique for brewery wastewater, could be a useful method for biodegradable wastewater utilization. It is an effective method for the removal of COD (Chemical Oxygen Demand), it exceeds 95%.

7.7 Aerobic Treatment

After primary treatment with anaerobic digestion, the effluent has high-quantity organic, which gives dark brown color to it. Hence, the anaerobic stage is followed by aerobic treatment, considered as secondary treatment. Some of the aerobic treatment methods are discussed below.

7.8 Aquaculture

In southern India, near Chennai city, farmers used post-methanated effluent for pisciculture. BOD (Biochemical Oxygen demand) is declined to zero after some time and yield of fish increased up to 50 tons per hectare per year (Billore et al. 2001).

7.9 Constructed Wetlands

Variable quantities of BOD (2500 mg/l) and COD (14,000 mg/l) were reported by Billore et al. (2001) after the analysis of post-anaerobic treated effluent. To reduce BOD and COD through aerobic procedures, the cell/container was filled with gravel up to variable levels of height; plants like *Typha latipholia* and *Phragmites karka* were grown, respectively. A time period of 14 days was provided and then result was recorded in content reduction of 85% BOD, 64% COD, 79% phosphorus, and 42% total solids (Trivedy and Nakate 2000).

7.10 Role of Enzymes

The role of enzymes in decolonization of wastewater is significant. Paper mills, tanneries, alcohol distilleries, dye-making units, and textiles are the main industries that synthesize and release colored wastewaters. They not only affect the soil and water but also act as a barrier between the path of light to the bed of the marine system which ultimately results in the reduction of photosynthesis, by developing anaerobic conditions, which in turn cause the death of aquatic life in toxified water. The most appropriate choice of enzyme for industrial utilization is Lignin Peroxidases (LiP), Manganese-dependent Peroxidases (MnP), and laccase, these are major enzymes to degrade lignin (D'Souza et al. 2006).

Although the benefits of enzymatic wastewater treatment are very obvious, the major drawback in the use of enzymes is their high cost. Hence, if enzymes can be reused after thoroughly applying the standardized immobilization procedures, the cost can be controlled to some extent.

7.11 Bacterial Role

Bacterial role in the decolorization of distillery effluent has been reported often. Bacterial strains from sewage habitat were isolated and then cultured on concentrations of distillery waste from low to high variable gradually. These bacteria were able to minimize COD (80%) without aeration, during 4–5 days. The major remnants after treatment are biomass, carbon dioxide, and volatile acids. A bioreactor with air bubble column, containing microbes with sludge is used for modifying aerobically the winery wastewater. The bacterial strains cultured from the reactor were Pseudomonas and Bacillus, which were often obtained from colonized carriers. *Pseudomonas fluorescens* decolorizes melanoidin wastewater (MWW) up to 76–90% in non-sterile and sterile samples differently. The variation in percentage could be because of the fact that melanoidin is sensitive toward variable pHs and temperatures. *Oscillatoria boryna*, a marine cyan bacteria, have been reported to produce

H_2O_2, hydroxyl, perhydroxyl, and active oxygen radicals, after the degradation of melanoidin, is the cause of color removal from the effluent. Acetogenic bacterial strain was used for decolorization up to 76.4% under the highest suitable nutrients. Hence, this value remained only 7.3% in the absence of oxygen. Glucose and fructose, forms of sugar, are also demanded for the decolorization of MWWs. Acetogenic bacteria can decompose melanoidins by oxidation and color might be removed due to a sugar oxidase.

7.12 Algal Species

Nowadays, algae are getting more spotlight on being one of the most reliable trending sources of food, biofuel, oils, biomass, and many more by-products. The standards which make algae so advantageous are the options to cultivate it in many ways, according to products. Alga is always been a reliable source to produce oil from billions of years and the way is the uniqueness of natural and engineered processes. Here are some reasons to elaborate on the importance of algae as a source of fuel and other by-products.

7.12.1 Purify Wastewaters

Algae grow in municipal wastewaters (sewage), animal wastes, and some industrial effluents. While growing in these effluents, it enables them to be purified by utilizing their rich nutrients, simultaneously producing a huge mass suitable for biofuels production.

7.12.2 Algae Consume CO_2

Plants grow using sunlight, consume carbon dioxide (CO_2) releasing oxygen (O_2) for others to breathe. In the same way, algae require more CO_2 for more productivity, which is obtained from effluent sources such as power plants, factories, and other such sources.

7.12.3 Algae Can Have High Biofuel Yields

The storage forms of energy in algae are oils and carbohydrates, which, after coping with their high productivity rate, can produce up to 2,000–5,000 gallons of biofuels per acre per year.

7.12.4 Used for Feed and Food

Microalgae produce a high content of protein and oil, which can be utilized to get either biofuels or animal feeds, or both. In addition, microalgal biomass is already recommended as a dietary supplement for human health.

7.12.5 Energy Source

Algal remnants after oil extraction are dried and pelletized by applying pressure and then will be utilized as fuel which can be burnt in industrial sources to generate power. *Ulva* is one of the examples (weblink 14), has been investigated as a fuel for use in the *SOFT cycle*, (Solar Oxygen Fuel Turbine) (weblink 15).

7.12.6 Fast Growth

Algae growth rate is greater than other productive crops. The rate of multiplication is double in hours than in days, therefore can be harvested daily. It has the ability to produce a greater volume of biomass or fuel in very less time.

7.12.7 Useful Products

Plastics, chemical feedstocks, lubricants, fertilizers, and even cosmetics, there is a wide range of products for a small market too.

There are two kinds of algae depending on size, microalgae and macroalgae. Microalgae have size up to 0.4 mm in diameter such as cyanobacteria and diatoms, while macroalgae consist of seaweed. Microalgae are mostly preferred by researchers for oil production due to less complex structure, fast growth, and high oil content. However, now researchers are converting their focus to seaweeds for biofuel production; perhaps this is an easily available resource (Evgeni 2011).

Across the world, scientists are researching many species of algae to check their ability to produce oil in massive (weblink 16). Other algal species studied, include *Clostridium saccharoperbutylacetonicum, Prymnesium parvum, Euglena gracilis, Botryococcus braunii, Chlorella, Dunaliella tertiolecta, Gracilaria, Pleurochrysis carterae,* and *Sargassum.* Percentage of oil content in dry weight of microalgae is variable from species to species (Table 7.3).

Table 7.3 Oil percentage in dry weight of microalgae (weblink 17)

Sr. No.	Algal species	Percentage oil content in dry weight
1.	*Ankistrodesmus* TR-87	28–40
2.	*Botryococcus braunii*	29–75
3.	*Chlorella* sp.	29
4.	*Chlorella protothecoides*	15–55
5.	*Crypthecodinium cohnii*	20
6.	*Cyclotella* DI-	35: 42
7.	*Dunaliella tertiolecta*	36–42
8.	*Hantzschia* DI-	160: 66
9.	*Nannochloris*	6–63
10.	*Nannochloropsis*	431–68
11.	*Neochloris oleoabundans*	35–54
12.	*Nitzschia* TR-114	28–50
13.	*Phaeodactylum tricornutum*	31
14.	*Scenedesmus* TR-84	45
15.	Schizochytrium	50–77
16.	*Stichococcus*	9–59
17.	*Tetraselmis suecica*	15–32
18.	*Thalassiosira pseudonana*	21–31

7.13 Mechanism to Fix Carbon in Algae

CCM is a carbon dioxide-concentrating mechanism, a process in microorganisms like cyanobacteria and algae which are autotrophs and photosynthetic specifically. It helps them to acclimate in an environment which has a limited availability of carbon dioxide (CO_2). It facilitates to retain $C_{i,}$ i.e., CO_2 and/or HCO_3, higher in concentration inside the cell than outside (Figs. 7.2 and 7.3). Carbonic anhydrase is an enzyme that catalyzes the interconversion of carbon dioxide and hydrogen carbonate reversibly with a much faster rate as compared to without enzyme. Carbonic anhydrase converts HCO_3 to CO_2, and concentrates it as substrate in Rubisco (ribulose bisphosphate carboxylase–oxygenase), which ultimately optimizes photosynthesis process even when CO_2 is low in environment (Rubisco is a dual-nature enzyme which is able to use carbon dioxide and oxygen as substrates simultaneously, but it selects oxygen only in the absence of carbon dioxide). The efficiency of Rubisco is enhanced by its placement in structures like the carboxysome in cyanobacteria or the pyrenoid in algae. The charged hydrogen carbonate needs specific channels to cross organellar membranes and enter the cell. Inorganic carbon converts to hydrogen carbonate at higher pH levels. To prevent the escape of carbon dioxide from cell, it is preferably trapped in the form of hydrogen carbonate anion. This effective mechanism of carbon concentration makes the algae a preferred tool for biotechnology and especially in transgenic methods to get more production of biofuel and biomass (weblink 18).

7.14 Production/Cultivation Systems

A variety of algae cultivation techniques were developed and some are still under consideration, open ponds, closed photobioreactors, fermentation tanks, and hybrid systems. There are a lot of ways to culture algae and this provides a unique place to algae in commercialization point of view.

The main objective to select any system for algae cultivation is to get a significant quantity of algae to extract oil, fuel, chemicals, and other industrial by-products. The accuracy of system depends on the location and kind of production. Following are some systems which are currently in practice (weblink 19).

7.15 Open Pond Systems

Open pond systems are also called raceway ponds because it gives the appearance of a race track. It is one-foot-deep shallow pond covering the area of acres depending on size. It provides exposure to sunlight which is a compulsory component to increase biomass. Paddle wheels or other water moving devices are used to keep the algae in rotation.

A dual-phase harvesting process is designed for algae, each phase comprises specific properties depending on the kind and requirement of algae. A portion of water is removed on daily basis gradually. The biomass collected and used for further processing to get oil-based by-products and then remnants dried to use as fuel or animal feed.

These systems are developing in time, one of the examples is in southern New Mexico, a highly commercial valued open pond system is developed. It is a project of Sapphire Energy. This is able to produce millions of gallons of fuel yearly.

7.16 Enclosed Systems

It has a variety of shapes like tubular, bag, panels, or maybe a simple greenhouse. The size is varying from kind of algae to the quantity of growth required. These systems can be placed in different orientations such as vertical, horizontal, oblique, or continuous rotation. It has capacity to be accomplished with additional artificial lights to obtain boosted production. The prominent benefit to use this system is to get be escape from weed, zooplankton, etc., which invade the algae, generally in open pond system.

7.17 Heterotrophic Fermentation

Scientists adopted a new technique to grow algae by providing them an alternative mode of nutrition. As algae are autotrophs, if provide with sugar, algae utilize it in dark and convert it to other products accordingly, a heterotroph way to generate energy. Some industries pursue this approach. Solazyme a company located in San Francisco is one of the pioneer companies pursuing this mode of cultivation. They are producing tens of thousands of gallons of algae-based fuel to satisfy the needs of the US navy. The product is already tested and mixed with other petroleum to be used in different vehicles accordingly. They are also fulfilling the requirements of other companies in the field of chemicals, nutrition, and cosmetics by using remnants of algae biomass after oil extraction.

7.18 Mixed Systems

Ohio-based company known as Phycal is working on a new technique to get benefits of all the above varieties in combination rather than individually. It is using open ponds, photobioreactors, and fermentors at a time or in a sequence; with this strategy, they are able to get benefits from all the modes of cultivation. This technique is named as "hybrid."

7.19 Coordinated Systems

This kind of algae cultivation process is most suitable for effluents from variable sources. In this category, alga is grown in wastewater, from which it absorbs carbon dioxide and degrades unwanted poisonous substances. The cultivated algal biomass is then used to obtain methane, fuels, and fertilizer.

The focus in this paper is on waste product. This waste from any source, either water or air, is used as nutrient or as input. In Iowa, algae (bioprocess) are utilizing a closed photobioreactor, which is a large vertical column of photobioreactors. It not only gets light but also collects carbon dioxide from ethanol plants (first generation).

One of the known corporations, Accelegy, designed a recycling mechanism to capture carbon from sources such as coal production units. Once carbon moves through the photobioreactor containing densely grown algae, the algae are separated and mixed with other materials and applied as biofertilizer to enhance crop growth where it continuously collects carbon dioxide from atmosphere.

7.20 Simultaneous Separation Processes

Simultaneous separation of useful product is a kind of excretion by which beneficial compounds are separated in the medium, instead of purifying algae and then processing it. This technique is followed by the genetic modifications in algae. A collaboration of Exxon and San Diego-based Synthetic Genomics is working on this type of experimentation. It will not only be economically better but one will be able to get desired compound like ethanol, butanol, fatty acids, hydrocarbons, gaseous fuels, and many more useful products without complicated processing.

7.21 Techniques to Extract Biofuels from Algae

7.21.1 Two Extraction Methods

There are two techniques of extraction. One of them requires water solution to disintegrate algal cells, while for other, water solution is devoided and cells are disrupted by any mechanical or chemical method (weblink 20). Mostly, four wet extraction techniques are applied either separately or combining them in any pattern.

7.22 Conclusion

Acknowledging the efforts of the United Nations and admitting that the United Nations Framework Convention on Climate Change is the primary international, intergovernmental forum for negotiating the global response to climate change, the above-mentioned methods will add on the reducing dangerous effects and interfering level of global greenhouse gases. These methods are not only easy to adopt but also convert carbon to beneficial form of biofuel and biomass through algae.

References

Andersson V, Broberg S, Hackl R, Karlsson M, Berntsson T (2014) Algae-based biofuel production as part of an industrial cluster. Biomass Bioenerg 71:113–124

Bhale PV, Deshpande NV, Thombre SB (2009) Improving the low temperature properties of biodiesel fuel. Renew Energy 34:794–800
Billore SK, Singh N, Ram HK, Sharma JK, Singh VP, Nelson RM, Dass P (2001) Treatment of molasses based distillery effluent in a constructed wetland in central India. Water Sci Tech 44(11–12):441–448
Chisti Y (2007) Biodiesel from microalgae. Biotechnol Adv 25:294–306
D'Souza DT, Tiwari R, Sah AK, Raghukumar C (2006) Enhanced production of laccase by a marine fungus during treatment of coloured effluents and synthetic dyes. Enzyme Microb Technol 38:504–511
Delrue F, Álvarez-Díaz P, Fon-Sing S, Fleury G, Sassi JF (2016) The environmental biorefinery: using microalgae to remediate wastewater, a win-win paradigm. Energies 9:1–32
Demirbas A (2010) Use of algae as biofuel sources. Energy Convers Manage 51:2738–2749
Dragone G, Fernandes B, Vicente A, Teixeira J (2010) Third generation biofuels from Microalgae. Curr Res Technol Educ Top Appl Microbiol Microb Biotechnol 1355–1366
Evgeni Y (2011) Seaweed ulva photosynthesis and zero emissions power generation. Int J Energy Environ Engin 2(1):23–31
Fon Sing S, Isdepsky A, Borowitzka MA, Moheimani NR (2011) Production of biofuels from microalgae. Mitig Adapt Strateg Glob Chang 18:47–72
Harder R, Witsch VH (1942) Bericht über versuche zur fettsynthese mittels autotropher microorganism men. Forschungsdienst Sonderheft 16:270–275
Hu Q, Sommerfeld M, Jarvis E, Ghirardi M, Posewitz M, Seibert M et al (2008) Microalgal triacylglycerols as feedstocks for biofuel production: perspectives and Advances. Plant J 54:621–639
Jämsä M, Lynch F, Santana-Sánchez A, Laaksonen P, Zaitsev G, Solovchenko A, Allahverdiyeva Y (2017) Nutrient removal and biodiesel feedstock potential of green alga UHCC00027 grown in municipal wastewater under Nordic conditions. Algal Res 26:65–73
Kilian O, Benemann CS, Niyogi KK, Vick B (2011) High-efficiency homologous recombination in the oil-producing alga *Nannochloropsis* sp. Proc Natl Acad Sci USA 108:21265–21269
Kumar M, Sharma MP (2016) Selection of potential oils for biodiesel production. Renew Sust Energ Rev 56:1129–1138
Kwaka M, Won-Kun I, Parkb 1, Shina S, Koha H, Leea B, Jeonga B, Changa Y(2017) Improvement of biomass and lipid yield under stress conditions by using diploid strains of *Chlamydomonas reinhardtii*. Algal Res. 26:180–189
Lynch F, Santana-Sánchez A, Jämsä M, Sivonen K, Aro EM, Allahverdiyeva Y (2015) Screening native isolates of cyanobacteria and a green alga for integrated wastewater treatment, biomass accumulation and neutral lipid production. Algal Res 11:411–420
Mata TM, Martins AA, Caetano NS (2010) Microalgae for biodiesel production and other applications: a review. Renew Sust Energ Rev 14:217–232
Moroney JV, Ynalvez RA (2007) Proposed carbon dioxide concentrating mechanism in *Chlamydomonas reinhardtii*. Euk Cell Minirev 6(8):1251–1259
Najafi G, Ghobadian B, Yusuf FT (2011) Algae as a sustainable energy source for biofuel production in Iran: a case study. Renew Sustain Energy Rev 15(8):3870–3876
Paswan G, Nikhil K (2014) Biopurification of waste water through algae—a review 71–73
Rani K, Sridevi V, Rao RSV, Kumar KV, Harsha N (2013) Biological treatment of distillery waste water—an overview. Int J Gen Eng 2(4):15–24
Ribeiro N, Pinto AC, Quintella CM, da Rocha GO, Teixeira LSG, Guarieiro LLN et al (2007) The role of additives for diesel and diesel blended (ethanol or biodiesel) fuels: a review. Energy Fuel 21:2433–2445
Subashchandrabose SR, Ramakrishnan B, Megharaj M, Venkateswarlu K, Naidu R (2013) Mixotrophic cyanobacteria and microalgae as distinctive biological agents for organic pollutant degradation. Environ Int 51:59–72

Trentacoste EM, Shrestha RP, Smith SR, Gle C, Hartmann A C, Hildebrand M et al (2013) Metabolic engineering of lipid catabolism increases microalgal lipid accumulation without compromising growth. Proc Natl Acad Sci 1–110
Trivedy RK, Nakate SS (2000) Treatment of diluted distillery waste by constructed wetlands. Ind J Environ Protec 20(10):749–753
Wijffels RH, Barbosa MJ (2010) An outlook on microalgal biofuels. Science 329:796–799

Webliography

https://sustainabledevelopment.un.org/topics/atmosphere
https://sustainabledevelopment.un.org/sdg13
https://www.ipcc.ch/site/assets/uploads/2018/12/SR15_Talanoa_Hoesung_6.12.18.pdf
https://www.ipcc.ch/site/assets/uploads/2018/12/Jim-Skea-SBSTA.pdf)
Paris Agreement. https://treaties.un.org/pages/ViewDetails.aspx?src=TREATY&mtdsg_no=XXVII-7-d&chapter=27&lang=en
European Commission, Climate Actions: Energy. https://ec.europa.eu/clima/policies/international/paris_protocol/energy/index_en.htm
Zhang X (2015) Microalgae removal of CO_2 from flue gas, Clean Coal Technol Res Rep https://bookshop.iea-coal.org.uk/reports/ccc-250/83697
All-gas, Description of the project. https://www.all-gas.eu/Pages/DescriptionofProject.aspx
Marketwire News Release: National Renewable Energy Laboratory Report. https://www.greenfuelonline.com
GreenFuel Web Site https://www.greencarcongress.com/2006/06/petrosun_drilli.html
Aquaflow Bionomic Corporation Web Site: https://www.aquaflowgroup.com/Cleantech.com
Knothe G, Van Gerpen JH, Krahl JJ, Van Gerpen JH (2005). The Biodiesel Handbook. https://doi.org/10.1201/9781439822357
https://pdfs.semanticscholar.org/2d54/0358470e557d92b1f0387d57c779bf2c3277.pdf
Lewis L (2005) Seaweed to breathe new life into fight against global warming. The Times Online, London
Seaweed Biofuels: Production of Biogas and Bioethanol from Brown Macroalgae (2012). https://Amazon.com
https://www.zeitmop.de/pdf/TOWARD_A_LIFE_SEA_NEAR_THE_DEAD_ONE.pdf
https://en.wikipedia.org/wiki/algae_fuel
Algal Carbon Dioxide Concentrating Mechanisms (2007) In: Book, James Moroney eLS (ed). https://doi.org/10.1002/9780470015902.a0000314
https://allaboutalgae.com
https://mickpeterson.org/Classes/Design/2009_10/Projects/Algae/
https://www.etipbioenergy.eu/markets-policies/financing-and-investment

Chapter 8
Sustainable Biogas Production from Distillery Wastewater

Abdullah Yasar, Sidra Mukhtar, and Amtul Bari Tabinda

Abstract The distillery industry has a main contribution to the world economy. This is also important contributors to environmental degradation with the release of a large amount of wastewater. The wastewater of distilleries consists of very high concentration of BOD, COD, total solids, phosphate, sulfate, and different hazardous compounds. It causes serious environmental dilemma. A variety of physicochemicals and biological processes have been also researched for the safety of environment, human beings, and other living organism's health. The prospective of biogas production from anaerobic digestion treatment process of distillery spent wash was also investigated. The anaerobic digestion needs less capital and production cost per unit in comparison to other renewable resources of energy, for example, hydro, sunlight, and wind. In addition, the potential of biogas toward electricity production is very economical. In addition, Sustainable Development Goals (SDGs) were achieved through biogas production. In the SDGs, nonetheless, the goal number seven initiates the need to employ efficient and sustainable renewable resources of energy to accomplish several global goals. The different precise objectives within SDG number 7 are such as considerably amplify the role of renewable energy sources to global energy matrix by 2030, boost up in the supply of contemporary and sustainable energy applications for all in emergent countries by expansion in infrastructure, enhance the energy efficiency improvement rate throughout the globe and technology up gradation. Biofuels from wastewater such as distilleries become the most advanced and efficient alternative energy source. These sustainable biofuels will be expected to play an important role in the quest of these above-mentioned goals. Energy is highly related with all of the other sustainable development goals, and recognition of these interconnections is significant to successfully execute them. It is now time to act in response to a historic agreement toward climate change solution in place. So the delivering on SDG7 is a foundation stone of that goal. On the other hand, if biofuels are anticipated to be fundamental in the recreation of sustainable development renewable energy goals in the discontinuation of fossil fuels generally on the basis of its environmental responsiveness, so this is the only prudent to evaluate how it completely merits its contemplation. However, the breakthroughs accomplished

A. Yasar (✉) · S. Mukhtar · A. B. Tabinda
Sustainable Development Study Center, GC University, Lahore, Pakistan
e-mail: yasar.abdullah@gmail.com

M. Arshad (ed.), *Sustainable Ethanol and Climate Change*,
https://doi.org/10.1007/978-3-030-59280-6_8

hitherto indicate that the future seems definitely bright and intense along with their power to generate a practicable neutral renewable energy resource from waste matter to sustain the comprehensive force toward the achievement of SDG number seven.

Keywords Distillery spent wash · Characterization · Treatment methods · Biogas production · Electricity generation · Sustainable development goals (SDGs)

- The wastewater from distillery industries has very deleterious effects on the surrounding environment.
- The prospective of biogas production from anaerobic digestion treatment process of distillery spent wash is innovative and effective in pollution reduction.
- Biogas potential toward electricity generation is very economical and environment friendly.
- It is extremely related to SDGs 7 due to the employment of efficient and renewable energy resources for the accomplishment of several energy goals.

8.1 Introduction

The rapid expansion of industrialization and development of civilization has drawn particular attention (Fito et al. 2017) to the inappropriate handling of waste materials that resulted in environmental deterioration (Narain et al. 2012). The energy consumption is growing at an alarming rate worldwide, resulting in depletion of fossil fuels (Arshad et al. 2014). Today, our environment is constantly destroying due to the discarding of industrialized waste matter into the surrounding environment (Nath et al. 2007). Hence, industrial pollution is the most important aspect that causes the degradation of the environment which is affecting the whole system (Saranraj 2012). Industrialized waste management is one of the major dilemma faced by today's world as manufacturing processes generate huge amount of wastewater contaminants that are complex to handle and costly to treat (Kadam and Upadhyay 2012). Moreover, the demand for water and the production of effluents is increasing at serious rate. The removal of wastewater is still a major crisis faced by many industries due to the complication and enormous volume of wastewater with very limited space for effluent treatment and finally disposal (Kumar and Chopra 2012) (Fig. 8.1).

The distillery industries are influential participators to the economy of world. These manufacturing mills are also regarded as the key resource of environmental contamination globally. About 319 distillery mills with 3.25×109 L of alcohol generation and production of wastewater is approximately 40.4×1010 L per year in India (Chandra et al. 2012). In 2007, the worldwide generation of bioethanol was about 50 billion liters while it arrived at 60 billion liters in lieu of about 4% of the gasoline usage globally in 2008 (Mussatto et al. 2010). Sugar molasses are utilized on a large level to produce ethanol. The ethanol generation during 2015 was approximately 224 crore L globally. With the generation of ethanol, around

Fig. 8.1 Industrialization

2,688 crore L of distillery wastewater was generated, which influences terrestrial and marine ecosystem very defectively (Wagh and Nemade 2018).

The distillery production mills utilize molasses, sugarcane, cereals, and additional agriculture goods for making alcoholic beverages. Sugarcane molasses is the major feedstock for distilleries in Pakistan for ethanol production (Arshad et al. 2017). The manufacture of fermented and refined beverages all over the world depends on substances that can be produced locally and is highly suitable to existing climatic situations. Ethyl alcohol is produced through fermenting molasses in the distillery industry. Molasses are produced from sugar making plants, which may be dependent on beet sugar or sugarcane. The agitation of molasses is done through yeast after appropriate dilution. Production of ethanol and byproducts from molasses-based media has been reported on laboratory scale (Arshad et al. 2008). The surplus end product of distillation is termed as spent wash or stillage and or distillery wastewater. Spent wash or stillage are one of the highly complicated, problem creating, and heavy industrialized organic wastewater. The toxic potency is very high because of the huge components of biologically degradable crude substances. For instance, the waste effluents have several organic compounds like dextrins, lignins, sugar, hemicelluloses, organic acids, and resins. Sugarcane molasses produced by sugar mills in Pakistan are stored in very poor conditions and so when they reach the distilleries, they are found to be loaded with contaminating microorganisms (Arshad et al. 2011).

Severe ecotoxicological and health-related hazards are produced when raw or incompletely treated distillery wastewater released into the surrounding environment. In the aquatic environment, the released effluents decrease the penetration strength of sunlight creating depletion in photosynthesis system and reduction in the concentration of dissolved oxygen (Saranraj and Stella 2014). While in soil mechanism, the productiveness of agricultural soil was reduced. Distillery wastewater should be sufficiently treated for the beak down of natural and inert contaminants former to its ultimate release into the surrounding environment because of these serious environmental and health hazards (Chowdhary et al. 2018).

Cost-effective and well-organized treatment techniques for distillery wastewater or spent wash are still in examination although the government policies which are controlling contamination are highly strict. Traditional effluent treatment methods typically comprise primary treatment to eliminate different materials such as fats, solids, and oils. The biologically secondary treatment is applied to eliminate organic materials and nutrients. The tertiary treatment of distillery wastewater may be known as polishing phase (Andrade et al. 2014).

Distillery industries for generation of several products and by-products such as biofuel, heat, and electricity have been in focuse in the recent years (Zhang 2008). In a biorefinery, biomass can be converted to useful materials in an integrated manner and thereby it can exploit the economic value of the biomass used while reducing the waste streams produced (Thomsen 2005). Development of multiple biofuel-based biorefinery from lignocellulose is seen as an important possibility to increase the efficiency of materials and energy. It removes the costs of biomass options to alleviate greenhouse gas emissions (Sheehan et al. 2003).

Biogas is developed as a primary treatment alternative from highly strong untreated spent wash during the complete scale use of a biomethanation method consisting of anaerobic fixed film reactors. The production of biomanure is done by the biocomposting of another surplus product of distillery industry like press mud. In the current circumstances, due to the generation of biogas anaerobic digestion is achieving broader acceptance above aerobic treatment process.

Biogas is renewable resource of energy with high efficient fuel characteristics and can be used for diverse energy services for instance fuel for transportation, electricity production, and other heat-related sectors (Ohimain and Izah 2017). The consumption of biogas would lessen the exploitation of fossil ful energy and decrease environmental threats (Mshandete and Parawira 2009). These impacts include pollution, global warming, and reducing demand for wood, advance sanitation, and reduction in charcoal demand for cooking (Paepatung et al. 2009). Biological conversion of waste suggests biological conversion of waste matter and the improvement of intricate organic raw waste into an important metabolite (Aliyu 2012). This conversion is performed in anaerobic digester. It utilizes microorganisms in anoxigenic situations to optimize the organic materials through converting it into methane and supplementary inorganic matters (Bambang et al. 2012). Numerous microbial species have been recognized for their capability to degrade organic matters present in waste effluents thus making value-added goods (Alam et al. 2007).

8.2 Characteristics of Distillery Wastewater or Spent Wash

The distillery wastewater is termed as spent wash. It is characterized by several qualities such as low pH, elevated temperature, high organic and inorganic substances, ash matter, and color. As an end result of the distillation and fermentation of molasses, spent wash is generated. The nature of spent wash is acidic. It has very high concentrations of BOD and COD 40,000–50,000 mg/l and 10,000–125,000 mg/l, respectively. About 90–93% of water, 7–10% of solids, 10–11% of proteins, and 2–20% of sugar are present in dry spent wash. The following metals are present in distillery wastewater like 348 mg/ltr Fe, 12.7 mg/ltr Mn, 3.65 mg/ltr Cu, 4.61 mg/ltr Zn, 0.64 mg/ltr Cr, 0.08 mg/ltr Co, and 0.48 mg/ltr Cd having 15–23 dS/m electrical conductivity range. The high concentrations of pollutant of spent wash are mostly due to the occurrence of high amounts of organic substances (Chowdhary et al. 2017).

Melanoidins are the main dye contaminants producing severe environmental hazards and health problems in human beings and other living organisms (Tamanna and Mahmood 2015). These are obstinate composites of sugar and amino acids and produced through the manufacturing of sugarcane juice in sugar industries and molasses in distillery factories (Arimi et al. 2015a, b; Saranraj and Stella 2014; Onyango et al. 2012; Wang et al. 2011).

8.3 Ecotoxicity and Health Risks Related to Distillery Spent Wash

Distilleries generate a high concentration of toxic compounds during different processes. These phenolic compounds consist of many substances like anticarcinogenic, antioxidant, antimicrobial characteristics (Silván et al. 2006). During beer storage, proteins react with these compounds and produce highly poisonous compounds and vapors (Siqueira et al. 2011).

Severe water pollution is caused by release of distillery wastewater when discharged into water resources without proper treatment. Eutrophication of polluted water bodies occurred due to its high nutrient content (Mahimaraja and Bolan 2004). LC50 charge of 0.5% was estimated through utilizing a biological toxicity experiment on aquatic fish named as Cyprinus carpio. Consequently, respiratory mechanism in Cyprinus carpio under high distillery effluent stress moves toward the anaerobic situations at organ stage in the hazardous conditions (Ramakritinan et al. 2005).

Due to inappropriate land discharge of distillery wastewater, acidification and soil pollution occurred. It restrains seed germination, causes nutrient deficiency, slows the growth rate, reduces crop yield, and decreases soil alkalinity (Chowdhary et al. 2017; Onyango et al. 2012).

In addition to soil and water pollution, serious health issues, for example, eyes irritation, stomach problem, fever; skin allergies, vomiting, etc., are faced by the inhabitants of distillery wastewater polluted area. The major reason for these problems is the occurrence of high amounts of various dissolved substances such as potassium, sodium, iron, magnesium, calcium chloride, carbonates, and bicarbonates (Chaudhary and Arora 2011).

Melanoidins observe in distillery spent wash have cytotoxic, carcinogenic, and genetic variation impacts on cells (Silván et al. 2006). Extreme glycation procedure also breaks down the necessary amino acids and stimulates the inactivation of enzymes. The disturbance in nucleic acid function, enhanced immunogenicity, and endocytosis also occurred (Taylor et al. 2004). Additionally, melanoidins were also engrossed in the development of different diseases for instance cardiovascular problems, Alzheimer's disease, and diabetes mellitus (Somoza 2005).

8.4 Treatment Procedures for Distillery Wastewater or Spent Wash

The majority of the distillery industries are experiencing environmental concerns related to treatment and disposal of distillery wastewater. A novel approach is needed which involves new processes and materials for effluent treatment.

8.4.1 Coagulation and Flocculation

The deterioration of colloidal elements through counterbalance the powers neutralization that remain them separately through applying coagulants are called as coagulation. The different coagulants used during the distillery wastewater treatment are like alum, aluminum sulfate, polyaluminum chloride, ferric chloride, iron aluminum, ferrous sulfate, calcium salts, etc. These coagulants are found to be effective to remove the organic and inorganic compounds from distillery spent wash (Wagh and Nemade 2015; Pandey et al. 2003).

Specific pH is necessary for specific coagulants and their efficiency is based upon many factors like type, amount, and properties of wastewater to be treated. The coagulants such as PACl, FeCl3, and AlCl3 have resulted in 72.5, 55, and 60% COD decrease and 92, 8 3, and 86% color removal from distillery wastewater (Chaudhari et al. 2005). Sowmeyan and Swaminathan (2008) have reported that the effectiveness of FeCl3 and AlCl3 for distillery wastewater treatment by accounting 93 and 76% removal in color and TOC, correspondingly. In addition, the highest color elimination approximately up to 98% was observed under alkaline environment (Pandey et al. 2003). Furthermore, Prajapati and Chaudhari (2015) have presented 92% COD removal and 81.8% color removal by using various coagulants. A green technology applying seed extract of Moringa oleifera like the coagulant in combination with chemical coagulants reported in 97% color removal (Fernandes et al. 2015).

8.4.2 Adsorption

Surface depended physical process exploited for the reduction of organic contaminants from industrialized effluents is termed as adsorption. Different absorbents are used during distillery wastewater treatment such as powdered activated carbon, chitosan, biochar, bagasse, etc. (Mandal et al. 2003). Activated carbon is commonly used for the reduction of color and particular organic contaminants from different industrialized effluents (Prajapati and Chaudhari 2015). Commercial activated charcoal treated the distillery wastewater and resulted in greater than 90–99% color and other pollutants removal (Chandra and Pandey 2000).

About 98 and 99% removal was observed in color and COD, correspondingly, by using chitosan as an absorbent for distillery wastewater treatment (Lalov et al. 2000). Additionally, about 50% color removal from distillery spent wash through utilizing chemically treated bagasse was accounted (Mane et al. 2006). Other researchers have also reported reduction in color and organic matters by 62.83% and 95.4%, respectively, by treating with bagasse activated carbon (Shivayogimath and Inani 2014).

8.4.3 Anaerobic Treatment

The anaerobic treatment is the mainly suitable technique for the alleviation of elevated organic carbon matter like distilleries. The anaerobic digestion procedure is primarily used to generate biogas from spent wash. The high amount organic substance of molasses spent wash forms the anaerobic digestion more effective in contrast with aerobic treatment method (Mohana et al. 2007). The organic complexes that occur in distillery wastewater are degraded with microorganisms to generate biogas. On a standard basis, 1 m^3 of spent wash generates approximately 38–40 m^3 of biogas. The supplementary products of anaerobic digestion comprise processed spent wash and degraded sludge, which is very loaded in nutrients. The most broadly used anaerobic procedure for distillery wastewater treatment process is UASB (Satyawali and Balakrishnan 2008).

Application of anaerobic digestion process for the treatment of distillery effluents is the most suitable primary treatment preference. Anaerobic treatment of wastewater in up-flow anaerobic sludge blanket system could be efficient process of simply biodegradable wastewater exploitation. The attained efficiency of pollutants removal was adequate as it arrived at above 95%. Another two similar UASB reactors worked in corresponding and 90% pollutant removal was achieved. A mesophilic two-phase method having an anaerobic filter and an up-flow anaerobic sludge blanket reactor was found appropriate for distillery wastewater anaerobic digestion (Rani et al. 2013). The performance efficiency of combined anaerobic digestion and ultraviolet process of highly distillery wastewater was very significant. About 88% reductions in pollutants was achieved (Apollo et al. 2013).

8.4.4 UASB Reactors

In recent years, UASB reactors have achieved more popularity for the treatment of distillery wastewater (Petta et al. 2017). It contains four major components. These components are gas–solid separator, sludge bed, settlement compartment, and sludge blanket. The process of UASB is generally based on the development of active and settled granules. These granules increase the effectiveness of UASB reactors (Akunna and Clark 2000). Several factors are involved in the performance and efficiency of USAB reactors such as pH, temperature, wastewater characterization, and organic loading rate.

The UASB reactors coupled with the UF-MBR acquire the process efficiency greater than 97% with the 340 L methane generation (Petta et al. 2017). Hence, the most important qualities of UASB reactor configuration consist of mechanical digester matters, sludge biomass recycling due to temperature and high organic loading (Sharma and Singh 2000). Another study described that USAB reactors can attain 90% COD reduction from distillery wastewater (Wolmarans and De Villiers 2002). In addition, biogas is also generated from the UASB treatment of distillery wastewater (Patyal 2016).

8.4.5 Aerobic Treatment

The post-anaerobic treatment stage still has high organic loading and is high dark brown in color, hence it is generally followed by a secondary, aerobic treatment.

8.4.6 Bacterial Treatment

These treatments utilizing uncontaminated bacterial growth have been studied and used normally in the history and current years. Degradation through bacteria and removal of color from industrial effluents is an environmental friendly and cost-effective preference to the physical and chemical treatment methods of effluents. The bacterial medium consisting of *Stenotrophomonas matophila* and *Pseudomonas aeruginosa* is assessed and about 67 and 51% removal efficiency in color and COD was obtained during 24 and 72 h, correspondingly at 37 °C from distillery effluents (Mohana et al. 2007).

8.4.7 *Mycoremediation*

Different fungal species have been applied during different processes for distillery effluent treatment (Bezuneh 2016). *Cladosporium cladosporioides* has the ability to remove 52.6 and 62.5% of color and COD, respectively, from distillery effluents (Ravikumar et al. 2011). In addition, *Cladosporium cladosporioides* is used at special circumstances such as fructose amount is 7 g per liter, peptone concentration is 2 g, pH 6, and the obtained results are higher than the previous (Ravikumar et al. 2013). Futher, more color removal efficiency was attained through the treatment of distillery industry wastewater through utilizing *Aspergillus niger* with the assistance of mixed coagulants was discussed (Shukla et al. 2014). Nevertheless, several white fungi are also studied to secret ligninolytic which are competent in destroying toxins and various organic and inorganic contaminants (Chandra and Chowdhary 2015).

8.4.8 *Phycoremediation*

The treatment of distillery wastewater through microalgae not only treats the waste but also its byproducts are very important for social well-being (Sankaran et al. 2014). The potential of distillery wastewater treatment through an emergent *Chlorella sorokiniana* species produced in a photobioreactor which has high density were studied (Solovchenko et al. 2014). The treatment through algal species becomes more efficient after the anaerobic treatment process of spent wash. The treatment process is energy competent and has the capacity to accomplish its nutrients necessity from biologically methanated spent wash and sunlight produces adequate energy. The anaerobically treated distillery wastewater is further treated through the microalgae *Chlorella vulgaris* chased via *Lemna minuscula* reported in 52% color removal efficiency (Valderrama et al. 2002). In addition, the destruction of 5% melanoidin through a marine *cyanobacterium* was observed (Kalavathi et al. 2001). Therefore, combining microalgae biomass generation with nutrient reduction and contaminant destruction may signify an essential goal in the bioenergy production purpose because the waste effluents market is huge (Sankaran et al. 2014).

8.5 Production of Biogas

Biogas produced during the treatment of distillery effluents in Up-flow Anaerobic Sludge Bioreactor (UASB) method. The reduction efficiency for COD is 60.43%. This is acceptable limit for anaerobic treatment of distillery wastewater through the UASB reactor. The standard temperature required for efficient treatment is in the limit of 35–40 °C. The optimal temperature of reactor was sustained at 37.58 °C with assistance of light bulbs. The methane gas generation has been attained 57%

and other gases 43% and if accurately controlled. It could characterize a possible resource of energy for heating and electricity generation. Chemical oxygen demand effects the generation of biogas in such a way that COD reduction rate goes on increasing then the biogas generation rate goes on rising (Patyal 2016).

Approximately 300 distilleries are working in India which produces about 3 billion L of alcohol. Annually 40 billions of wastewater were generated. These alcoholic distilleries have great potential to generate 1300 million cubic meters of efficient biogas (Kumar and Aklilu 2016).

Production of biogas from the biological waste in anaerobic digestion is a renewable energy source. Biogas contains 50–60% methane (CH_4), 30–40% carbon dioxide (CO_2), moisture, and the traces of Hydrogen sulfide (H_2S) due to the presence of CO_2 and H_2S the working of the engine was affected badly. In this study, a method of biogas scrubbing and methane enrichment was presented. An experiment was established in which the chemical absorption of CO_2 and H_2S by aqueous solution was studied.

Since the last 50 years, Brima Sagar Maharashtra Distilleries Limited has been generating and distributing alcohol and other products. Their production plant employs molasses to produce alcohol. The effluents released from distillery were treated by biomethanation process to generate biogas.

The aqueous solutions used are sodium hydroxide, calcium hydroxide, and monoethanolamine. After the liquid aqueous solution circulation the absorbent characteristics were examined. And the test results explained that the aqueous solutions are more effective in reaction with CO_2 in biogas. This technique was approved effective in improving the biogas quality (Tippayawong and Thanompongchart 2010).

In India, generation of biogas from anaerobic digestion of biomass was demonstrated. It is the most efficient and effective option in other alternative sources of energy as wind, water, and solar. Renewable energy is also available as a domestic source in rural areas. In 2006, in India, there was a deficit of 11, 436 MW equivalent to 12.6% which is due to demand from various energy sectors increased and the energy supply was not in place with the demand till 2007 the total installed capacity of bioenergy generation from solid biomass and waste to energy is about 12,227 MW against potential of 25,700 MW. The bioenergy potential from various sectors as dairy plants, pulp and paper mill, poultry and sugar industry was estimated to be 40.734 Nm^3/year. (Rao et al. 2010).

The yield of biogas and their energy potential in Brazil through steam-exploded straw and bagasse was assessed. The steam explosion was the main type of thermal pretreatment. The process occurs under different conditions of temperature and time. This biogas yield was analyzed in batch process. The substrate used was bagasse and sugarcane straw. Earlier the biogas straw was burnt in the fields and now it was used for biogas production. The results of this study revealed that pretreatment of steam increases the production capacity of biogas from bagasse and the cane straw in comparison with untreated water. Due to the utilization of these wastes in Brazil, it would have high independence in energy sector because Brazil has higher usage of ethanol in transportation zone. The application of this technology on huge scale plants was also under consideration (De Paoli et al. 2011).

Ethanol process as a pretreatment for the biogas production under the laboratory conditions was studied. The pretreated oat straw was fermented to ethanol and the waste after ethanol removal was used for the biogas production. The material was grown in Sweden; the pretreatment used was thermo-chemical treatment. The detailed analysis shows that the combination of biogas digestion and the ethanol fermentation increases 85–87% conversion of halocellulose into biofuel energy. CO_2 formed during ethanol production effects the process but if this effect is small then it was avoided. The energy of both ethanol and methane was higher as compared to the methane alone. The rate of biogas production was also higher from the fermentation of distillery waste (Dererie et al. 2011).

Due to the production of the biogas through the anaerobic process, it becomes more popular than aerobic treatment which can also be used for meeting the energy crisis in the country. In this paper, some examples of the anaerobic digestion of the effluents of the industrial waste were discussed. The industrial sectors which have maximum capacity of energy production were paper mill and the other sector was the sugar distillery which produces electric energy. The analysis includes the nature of the wastewater coming from different areas of industries. The technologies and the policies were evaluated in India for large scale biogas plants and energy production. This was a questionnaire based study (Lata et al. 2002).

The production of biogas through the animal manure and the waste slurries was studied under the process of anaerobic digestion. According to European countries, the ordinances they formed formulate a target that till 2020 they acquire 20% of energy demand through renewable energy resources. The primary treatment of waste matter and animal dung because of biogas generation becomes an environmental friendly and sustainable resource. As a result of emergence of biogas in the global market, numerous issues of economy can be resolved. The major advantage of biogas was its instant storage in comparison to others and can be utilized easily in substitute for the natural gas (Holm-Nielsen et al. 2009).

8.6 Electricity Production from Biogas

Biogas Support Program was started by Pakistan government in 2000, up till now it has achieved the target of installing 12,000 units and in the next 5 years another 10,000 units are expected to install. And it will fulfill 27% of country biogas demand. A biogas unit of 10 m^3 is anticipated to save almost 92.062 PKR/year on account of conventional fuel. These biogas generation system is in demand and they are low cost and can be run with very small budget. There is a need of National policy to bring this new technology (Amjid et al. 2011).

A process for wastewater treatment of industry which produces electricity through biogas was developed. The intermediate between wastewater and the electricity production was biogas-fueled fuel cells that have an important impact on the process.

These fuel cells are widely used in applications but there construction leads to the fueling of biogas which can be used for the feeding of biogas for energy production. The biological process is applicable where the wastewater contains higher biological demand and chemical oxygen demand, so this study focuses on the biological processes (Yentekakis et al. 2008).

Lansing stated that methane, electricity production and the wastewater modifications in a digestion method. They examined the combined capacity of dairy and swine collectors in Costa Rica. On a comparison of study, dairy digesters generated 27.5 m^3/day biogas and swine reactors generated 6.0 m^3/day and the COD removal was 86% and 92%, respectively. These two combined digesters increased the electricity production because of effective generation capability of dairy and swine reactors. The electricity demand was accomplished to about 81% manure of 40 cows' remains in corrals for 100% used for electricity generation. Due to lowering wastewater and higher electricity which save the cost of project was investigated (Lansing et al. 2008).

Nguyen suggests that the major benefits of bioenergy system reduced greenhouse gas emissions. In comparison to fossil fuel, converting a sugar industry into a bioenergy plant would reduce climate variation by the production of electricity and ethanol. This study takes the example of Thailand sugar industry. Production of ethanol from surplus sugar has good saving potential of 14 million tons co_2/year as compared to the current practice of the sugar industry. According to this study, in Thailand, it takes 4.5–7 years which convert the land into a green land (Nguyen et al. 2010). The electricity produced from biomass is of favorable price and commonly available but has high land and water usage as well as social impacts (Evans et al. 2010).

As the country modernize energy utilization for household and for generation of new energy increased instead of minimizing utilization of bioenergy. The most obvious remains for biogas generation are bagasse, rice husk, and oil palm, due to their accessibility in the industries, and because of this bioenergy production system is feasible. Industrial wastewater and livestock dung have a possibility of 7800 and 13,000 TJ. New approaches were developed which overcome the energy utilization in Thailand (Prasertsan and Sajjakulnukit 2006).

Economical and the technological analysis of the generators producing energy with biogas coming from anaerobic digestion were documented. The literature revealed that first generator used for the production of electricity was set in Laraklio city in Greece for the management of industrial waste. The study includes a time period of 5.5 years. They stated that if only one generator becomes operational it produced less electricity, if used two the biogas used was increased and the electricity production efficiency also increased. So, almost six generators were used for this process. This process is economically very feasible as its cost is very less as compared to conventional process (Tsagarkis 2007).

Two sugar mills in Nicaragua decided for the generation of electricity from the bagasse in the sugar period and the other portion from the eucalyptus and sold it to national power station to get the desired profits. It was stated that the electricity generated from the sugar waste material should prove cheaper than the fuel in oil. In Nicaragua, 64% of the money which is used for electricity generation remains these otherwise 83% goes to other countries in case of fuel oil which was a conventional method of electricity production. The optimal employment for the electricity production from biomass should be higher than the fuel depended. In this study, a comparison was made between biomass-based and the fuel-based electricity and the biomass-based electricity proves good (van den Broek et al. 2000).

8.7 Role of Biogas Toward Sustainable Development

Sustainable development is defined as the development that meets the needs of the present generation without compromising the capacity of future generations to meet their own requirements. It is the main principle for maintaining limited sources essential to supply for the requirements of future generations (Leopold 2004). Under anaerobic conditions, biogas initiates from bacteria through the method of biodegradation of organic matters. An essential component of the biogeo-chemical carbon cycle is the natural production of biogas. In the sequence of microorganisms that breakdown organic matters and return the decayed products to the environment, methane generating bacteria are the last connection. A resource of renewable energy is produced through biochemical carbon cycle. It can be utilized for producing electricity and also as fuel in transportation sector. Biogas is also employed as an alternate for CNG (Zishan and Alayi 2013). The calorific value of biogas is approximately 6 kWh/m^3. Methane is the most important constituent in case the biogas is to be utilized in the form of fuel.

Biogas formation and utilization is a combined mechanism and participates in numerous sectors as shown in Fig. 8.2.

Energy is a fundamental element of society, environment, and economic development. Biogas plays a very important role in environmental sustainability. It contributes to the decrease of greenhouse gases, and preservation of forest. It facilitates in the improvement of health and hygiene by supplying clean energy and smoke-free kitchen which is directly related to the health of population and environmental protection. Health and environment alongside with pleasant atmosphere play a key role in better enterprise integration (Yadav 2014). The distinctive gas compositions are methane 55% and carbon dioxide 45% (D Vriens et al. 1985).

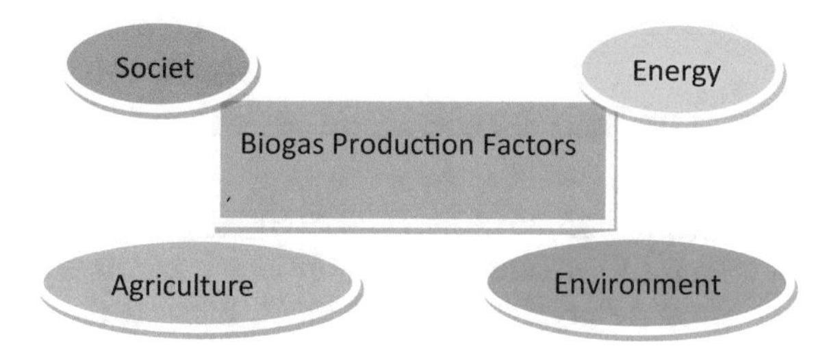

Fig. 8.2 Biogas production features

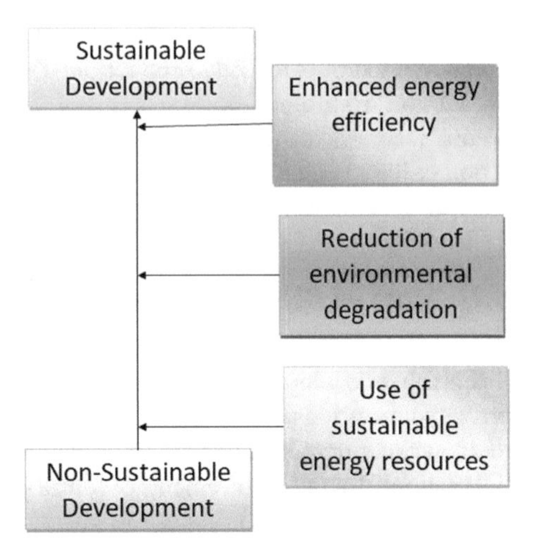

Fig. 8.3 Comparison of sustainable and non-sustainable development

During the last decades, anaerobic digestion has formed from a relatively simple method of biomass degradation. The efficient and better working biogas systems can produce a variety of advantages for consumers, civilization, and environment. The benefits are

1. Energy production
2. Organic wastes converted into best quality fertilizers
3. Enhancement of clean situations
4. Workload reduction
5. Positive environmental externalities
6. Economic advantages through energy and fertilizer replacement
7. Other economic and environment-friendly benefits through distributed energy production
8. Minimizes Greenhouse emissions effect.

Biogas technology considerably contributes to preservation and development. The figure shows the relationship between sustainable and non-sustainable development (Alayi et al. 2016) (Fig. 8.3).

8.8 Achievement of Sustainable Development Goals (SDGs) Through Biogas Production

On September 25, 2015, the United Nations (UN) adopted 17 aspirational Sustainable Development Goals (SDGs) and 169 targets as part of a global partnership to alleviate poverty and starvation in all its kinds, mitigate the effects of climate change, and ensure prosperous, fulfilling, and peaceful lives for all. The SDGs are targets to be met by 2030. This gives governments and businesses less than 13 years to make the urgent and necessary changes required to bring about and deliver a step change in sustainable development.

The biogas industry is uniquely positioned to help achieve nine of the SDGs as shown in Table 8.1. Anaerobic digestion is a natural method in which microorganisms degrade organic matter in sealed containers, producing biogas which can be used for cooking, heating, cooling, and electricity production or upgraded and used for vehicle fuel or gas-grid injection. This can be done on a microscale (for buildings or small communities) and on a macroscale (for cities). Biogas can also be naturally occurring, rather than produced in digesters, such as in Lake Kivu in Africa and in mines.

The fifth assessment report (AR5) of the United Nations' IPCC published in 2014 recommended that fossil fuels have to absolutely be phased out near 2100 (Climate change 2014). At this time, the report signifies that renewable energy resource composes up approximately 30% of world energy distribution and should rise to no less than 80% near 2050 if the 2100 goal is to be accomplished (Renewable energy statistics 2015).

According to COP24 report, the monitoring of progress in health and climate change is improving, but there are weaknesses in coverage and in stakeholder engagement. The indicators of the Sustainable Development Goals (SDGs) for climate change do not include health, although the situation is being remedied in academic research initiatives, by WHO and by the Secretariat of the UN Framework Convention on Climate Change (UNFCCC), in partnership with countries. Such indicators could be used for formal reporting to the UNFCCC, broader outreach to the public and monitoring of the achievement of the Sustainable Development Goals (SDGs) (WHO COP24 2018).

8.9 Conclusion

The distillery sector is the influential contributor to the world economics. This sector is also regarded as one of the key cause of environmental contamination globally. Severe ecotoxicological and health-related hazards are produced when raw or incompletely treated distillery wastewater is released into the surrounding environment. Cost-effectiveness and well-organized treatment techniques for wastewater are still in examination although the government laws which are controlling contamination are highly strict. Biogas is developed as a primary treatment alternative from highly strong untreated spent wash during the complete scale use of a biomethanation method consisting of anaerobic fixed film reactors. The production of biomanure is done by the biocomposting of another surplus product of distillery industry like press mud. In the current circumstances, due to the generation of biogas anaerobic digestion is achieving broader acceptance above aerobic treatment process. Environmental sustainability is enhanced through biogas production. It plays a very important role for removal of greenhouse gas emission and forest conservation. It assists in the improvement of health and sanitation through supplying clean energy and smokeless

Table 8.1 Achieving Sustainable Development Goals (SDGs) through biogas

SDGs	Biogas contribution (anaerobic digestion)
Goal 2: alleviate starvation, attain food security and improved nutrition and promote sustainable agriculture	• Restoring soils through the recycling of nutrients, organic matter, and carbon • Increasing crop yields through use of nutrient-rich digestate biofertiliser • Recirculating phosphorus, which is essential for the growth of plants but limited in supply
Goal 3: Make certain strong lives and endorse welfare for all at all ages	• Reducing indoor air pollution by substituting solid biomass-based domestic fuels with biogas • Treating and recycling sewage and organic wastes to reduce odors and the spread of diseases
Goal 5: Achieve gender equality and empower all women and girls	• Reducing the burden of collecting firewood to improve the quality of women's and children's lives, reducing household labor in cooking
Goal 6: Ensure accessibility and sustainable management of water and sanitation for every one	• Providing decentralized, local treatment of biosolids in remote and rural communities to reduce odors and the spread of disease • Stabilizing and recycling biosolids through AD to allow them to be applied back to land • Reducing the carbon loading of wastewater to reduce impact on water bodies
Goal 7: Make sure access to reasonable, reliable, sustainable, and efficient energy for every one	• Reducing dependence on fossil fuel-based energy sources by replacing with biogas • Capturing waste heat from co-generating units linked to biogas plants • Utilizing locally produced wastes and crops to generate energy for rural and remote communities • Storing biogas to produce energy when required
Goal 9: Build resilient infrastructure, promote inclusive and sustainable industrialization and foster innovation	• Improving the self-sufficiency and sustainability of industries by extracting the energy from their own effluents and using it for the self-generation of electricity and heat • Collaboration between industries and agriculture for mutual benefit • Generating short-term construction employment and long-term equipment manufacturing and maintenance employment • Encouraging growth of microenterprises by providing reliable electricity that can be stored and used when needed

(continued)

Table 8.1 (continued)

SDGs	Biogas contribution (anaerobic digestion)
Goal 11: Make cities and human settlements inclusive, safe, resilient, and sustainable	• Preventing spread of diseases through collection and proper management of organic waste • Improving sanitation and hygiene through decentralized and local treatment of biosolids • Stabilizing the sludge from wastewater treatment to protect the marine environment and urban air quality • Improving urban air quality by substituting fossil fuel with biomethane in vehicles • Improving urban air quality by substituting solid fuel for domestic cooking and heating with biogas • Reducing greenhouse gas emissions by using biogas-based renewable energy in buildings, homes, and industry
Goal 13: Take imperative action to contest climate change and its effects	• Reducing carbon dioxide emissions by replacing fossil fuel-based energy sources with biogas and commercial fertilizers with digestate biofertiliser • Reduction of methane and nitrous oxide emissions from livestock manures • Reduction of methane and generation of renewable energy from food and other organic wastes • Capturing emissions from landfills • Reducing deforestation by replacing solid biomass-based domestic fuels with biogas
Goal 15: Protect, restore, and promote sustainable use of terrestrial ecosystems, sustainably manage forests, combat desertification, and halt and reverse land degradation and halt biodiversity loss	• Recirculating nutrients and organic matter in organic wastes through AD and returning them to the soil in the form of digestate biofertiliser • Substituting firewood with biogas as a domestic fuel, reducing deforestation

kitchen which is directly related with the health of community and environment. Health and environment with eco-friendly environment participate for enhanced enterprise incorporation. The United Nations (UN) has proposed 17 Sustainable Development Goals (SDGs). The biogas technology is positioned to help achieve nine of the SDGs.

References

Adopted I (2014) Climate change 2014 synthesis report

Akunna JC, Clark M (2000) Performance of a granular-bed anaerobic baffled reactor (GRABBR) treating whisky distillery wastewater. Bioresour Technol 74(3):257–261

Alam MZ, Kabbashi NA, Razak AA (2007) Liquid state bioconversion of domestic wastewater sludge for bioethanol production. In: 3rd Kuala Lumpur international conference on biomedical engineering, vol 2006. Springer, Berlin Heidelberg, pp 479–482

Alayi R, Shamel A, Kasaeian A, Harasii H, Topchlar MA (2016) The role of biogas to sustainable development (aspects environmental, security and economic). J Chem Pharm Res 8(4):112–118

Aliyu S (2012) Palm oil mill effluent: a waste or a raw material? JASR 466–473

Amjid SS, Bilal MQ, Nazir MS, Hussain A (2011) Biogas, renewable energy resource for Pakistan. Renew Sustain Energy Reviews 15(6):2833–2837

Andrade LH, Mendes FDS, Espindola JC, Amaral MCS (2014) Nanofiltration as tertiary treatment for the reuse of dairy wastewater treated by membrane bioreactor. Sep Purif Technol 126:21–29

Apollo S, Onyango MS, Ochieng A (2013) An integrated anaerobic digestion and UV photocatalytic treatment of distillery wastewater. J Hazard Mater 261:435–442

Arimi MM, Zhang Y, Geißen SU (2015a) Color removal of melanoidin-richindustrial effluent by natural manganese oxides. Sep Purif Technol 150:286–291

Arimi MM, Zhang Y, Götz G, Geißen SU (2015b) Treatment of melanoidin wastewater by anaerobic digestion and coagulation. Environ Technol 36(19):2410–2418

Arshad M, Khan ZM, Shah FA, Rajoka MI (2008) Optimization of process variables for minimization of byproduct formation during fermentation of blackstrap molasses to ethanol at industrial scale. Lett Appl Microbiol 47(5):410–414

Arshad M, Zia MA, Asghar M, Bhatti H (2011) Improving bio-ethanol yield: using virginiamycin and sodium flouride at a Pakistani distillery. Afr J Biotechnol 10(53):11071

Arshad M, Ahmed S, Zia MA, Rajoka MI (2014) Kinetics and thermodynamics of ethanol production by Saccharomyces cerevisiae MLD10 using molasses. Appl Biochem Biotech 172(5):2455–2464

Arshad M, Hussain T, Iqbal M, Abbas M (2017) Enhanced ethanol production at commercial scale from molasses using high gravity technology by mutant S. cerevisiae. Braz J Microbiol 48(3):403–409

Bambang T, Vivian W, Yoshimasa T (2012) Methane emission from digestion of palm oil mill effluent (POME) in a thermophilic anaerobic reactor. Int J Sci Eng 3(1):32–35

Bezuneh TT (2016) The role of microorganisms in distillery wastewater treatment: a review. J Bioremediat Biodegradation 7(375):2

Chandra R, Chowdhary P (2015) Properties of bacterial laccases and their application in bioremediation of industrial wastes. Environ Sci Process Impact 17(2):326–342

Chandra R, Pandey PK (2000) Decolorization of anaerobically treated distillery effluent by activated charcoal adsorption method. Indian J Environ Prot 21(2):134–137

Chandra R, Bharagava RN, Kapley A, Purohit HJ (2012) Characterization of *Phargmites cummunis* rhizosphere bacterial communities and metabolic products during the two stage sequential treatment of post-methanated distillery effluent by bacteria and wetland plants. Bioresour Technol 103:78–86

Chaudhari PK, Mishra IM, Chand S (2005) Catalytic thermal treatment (catalytic thermolysis) of a biodigester effluent of an alcohol distillery plant. Ind Eng Chem Res 44(15):5518–5525

Chaudhary R, Arora M (2011) Study on distillery effluent: chemical analysis and impact on environment. Int Adv Eng Technol 2(2):352–356

Chowdhary P, Yadav A, Kaithwas G, Bharagava RN (2017) Distillery wastewater: a major source of environmental pollution and its biological treatment for environmental safety. In: Singh R, Kumar S (eds) Green technologies and environmental sustainability. Springer International, Switzerland, pp 409–435

Chowdhary P, Raj A, Bharagava RN (2018) Environmental pollution and health hazards from distillery wastewater and treatment approaches to combat the environmental threats: a review. Chemosphere 194:229–246
De Paoli F, Bauer A, Leonhartsberger C, Amon B, Amon T (2011) Utilization of by-products from ethanol production as substrate for biogas production. Bioresour Technol 102(11):6621–6624
Dererie DY, Trobro S, Momeni MH, Hansson H, Blomqvist J, Passoth V, Ståhlberg J (2011) Improved bio-energy yields via sequential ethanol fermentation and biogas digestion of steam exploded oat straw. Bioresour Technol 102(6):4449–4455
Eurostat (2015) Renewable energy statistics
Evans A, Strezov V, Evans TJ (2010) Sustainability considerations for electricity generation from biomass. Renew Sust Energy Rev 14(5):1419–1427
Fernandes DM, Sousa RM, de Oliveira A, Morais SA, Richter EM, Muñoz RA (2015) Moringa oleifera: A potential source for production of biodiesel and antioxidant additives. Fuel 146:75–80
Fito J, Tefera N, Van Hulle SW (2017) Adsorption of distillery spent wash on activated bagasse fly ash: kinetics and thermodynamics. J Environ Chem Eng 5(6):5381–5388
Holm-Nielsen JB, Al Seadi T, Oleskowicz-Popiel P (2009) The future of anaerobic digestion and biogas utilization. Bioresour Technol 100(22):5478–5484
https://www.clarke-energy.com/2014/brima-sagar-distillery-biogas-plant-2/
Kadam A, Upadhyay K (2012) Wastewater treatment of alcohol distillery. J Ind Poll Cont 28:1–4
Kalavathi DF, Uma L, Subramanian G (2001) Degradation and metabolization of the pigment—melanoidin in distillery effluent by the marine cyanobacterium Oscillatoria boryana BDU 92181. Enzyme Microb Technol 29(4):246–251
Kumar MR, Aklilu EG (2016) Production of biogas fuel from alcohol distillery plant. Int J Sci Res Publ 6(12)
Kumar V, Chopra AK (2012) Fertigation effect of distillery effluent on agronomical practices of Trigonella foenum-graecum L. (Fenugreek). Environ Monit Assess 184(3):1207–1219
Lalov IG, Guerginov II, Krysteva MA, Fartsov K (2000) Treatment of waste water from distilleries with chitosan. Water Res 34(5):1503–1506
Lansing S, Víquez J, Martínez H, Botero R, Martin J (2008) Quantifying electricity generation and waste transformations in a low-cost, plug-flow anaerobic digestion system. Ecol Eng 34(4):332–348
Lata K, Kansal A, Balakrishnan M, Rajeshwari KV, Kishore VVN (2002) Assessment of biomethanation potential of selected industrial organic effluents in India. Resour Conserv Recycl 35(3):147–161
Leopold A (2004) A sand county almanac. This Sacred Earth, Religion, nature, environment
Mahimaraja S, Bolan NS (2004) Problems and prospects of agricultural use of distillery spentwash in India. Super Soil 2004. 3rd Australian New Zealand Soils Conference, 5–9 December 2004. University of Sydney, Australia
Mandal A, Ojha K, Ghosh DN (2003) Removal of colour from distillery wastewater by different processes. Indian Chem Eng 45(4):264–267
Mane JD, Modi S, Nagawade S, Phadnis SP, Bhandari VM (2006) Treatment of spentwash using chemically modified bagasse and colour removal studies. Bioresour Technol 97(14):1752–1755
Mohana S, Desai C, Madamwar D (2007) Biodegradation and decolourization of anaerobically treated distillery spent wash by a novel bacterial consortium. Bioresour Technol 98(2):333–339
Mshandete AM, Parawira W (2009) Biogas technology research in selected sub-Saharan African countries—a review. Afr J Biotechnol 8(2)
Mussatto SI, Dragone G, Guimarães PM, Silva JPA, Carneiro LM, Roberto IC, Vicente., Domingues L, Teixeira JA, (2010) Technological trends, global market, and challenges of bio-ethanol production. Biotechnol Adv 28:817–830
Narain K, Yazdani T, Bhat MM, Yunus M (2012) Effect on physico-chemical and structural properties of soil amended with distillery effluent and ameliorated by cropping two cereal plant spp. Environ Earth Sci 66(3):977–984

Nath K, Singh D, Sharma YK (2007) Combinatorial effects of distillery and sugar factory effluents in crop plants. J Environ Biol 28(3):577
Nguyen TLT, Gheewala SH, Sagisaka M (2010) Greenhouse gas savings potential of sugar cane bio-energy systems. J Clean Prod 18(5):412–418
Ohimain EI, Izah SC (2017) A review of biogas production from palm oil mill effluents using different configurations of bioreactors. Renew Sust Energy Rev 70:242–253
Onyango MS, Ojijo VO, Ochieng A, Kittinya JO, Otieno FO (2012) Simultaneous adsorption and biodegradation of synthetic melanoidin. Afr J Biotechnol 11(22):6083–6090
Paepatung N, Nopharatana A, Songkasiri W (2009) Bio-methane potential of biological solid materials and agricultural wastes. Asian J Energy Environ 10(1):19–27
Pandey RA, Malhotra S, Tankhiwale A, Pande S, Pathe PP, Kaul SN (2003) Treatment of biologically treated distillery effluent-a case study. Int J Environ Stud 60(3):263–275
Patyal V (2016) Study of biogas generation in treatment of distillery wastewater by UASB method. Int J Eng Res Technol 5:634–638
Petta L, De Gisi S, Casella P, Farina R, Notarnicola M (2017) Evaluation of the treatability of a winery distillery (vinasse) wastewater by UASB, anoxic-aerobic UF-MBR and chemical precipitation/adsorption. J Environ Manage 201:177–189
Prajapati AK, Chaudhari PK (2015) Physicochemical treatment of distillery wastewater—a review. Chem Eng Commun 202(8):1098–1117
Prasertsan S, Sajjakulnukit B (2006) Biomass and biogas energy in Thailand: potential, opportunity and barriers. Renew Energy 31(5):599–610
Ramakritinan CM, Kumaraguru AK, Balasubramanian MP (2005) Impact of distillery effluent on carbohydrate metabolism of freshwater fish, Cyprinus carpio. Ecotoxicology 14(7):693–707
Rani K, Sridevi V, Rao RSV, Kumar KV, Harsha N (2013) Biological Treatment of Distillery Waste Water-An Overview||. Int J Gen Eng Technol 2(4):15–24
Rao PV, Baral SS, Dey R, Mutnuri S (2010) Biogas generation potential by anaerobic digestion for sustainable energy development in India. Renew Sust Energy Rev 14(7):2086–2094
Ravikumar R, Vasanthi NS, Saravanan K (2011) Single factorial experimental design for decolorizing anaerobically treated distillery spent wash using cladosporium cladosporioides. Int J Environ Sci Technol 8(1):97–106
Ravikumar R, Vasanthi NS, Saravanan K (2013) Biodegradation and decolorization of distillery spent wash with product release by a novel strain Cladosporium cladosporioides: optimization and biokinetics. CBEQ 27(3):373–383
Sankaran K, Premalatha M, Vijayasekaran M, Somasundaram VT (2014) DEPHY project: distillery wastewater treatment through anaerobic digestion and phycoremediation—a green industrial approach. Renew Sust Energy Rev 37:634–643
Saranraj P (2012) Effect of bacterial isolates on reduction of physico chemical characteristics in sugar mill effluent. IJPBA 3(5)
Saranraj P, Stella D (2014) Impact of sugar mill effluent to environment and bioremediation: a review. WASJ 30(3):299–316
Satyawali Y, Balakrishnan M (2008) Treatment of distillery effluent in a membrane bioreactor (MBR) equipped with mesh filter. Sep Purif Technol 63(2):278–286
Sharma J, Singh R (2000) Characterization of sludge from UASB reactors operating on molasses based distillery effluent. Indian J Microbiol 40(3):203–206
Sheehan J, Aden A, Paustian K, Killian K, Brenner J, Walsh M, Nelson R (2003) Energy and environmental aspects of using corn stover for fuel ethanol. J Ind Eco 7(3–4):117–146
Shivayogimath C, Inani S (2014) Treatment of biomethanated distillery spent wash by adsorption process on bagasse activated carbon. Int J Appl Sci Eng Res 3:1069–1078
Shukla SK, Tripathi A, Mishra PK (2014) Fungal decolorization of Anaerobically Biodigested Distillery Effluent (ABDE) following coagulant pretreatment. Int J Sci Environ Technol 3(2):723–734

Silván JM, van de Lagemaat J, Olano A, Del Castillo MD (2006) Analysis and biological properties of amino acid derivates formed by Maillard reaction in foods. J Pharma Biomed Anal 41(5):1543–1551

Siqueira PB, Maria H, Macedo GA (2011) Polyphenols and antioxidant properties in forced and naturally aged Brazilian beer. JBD 2(3):44–49

Solovchenko A, Pogosyan S, Chivkunova O, Selyakh I, Semenova L, Voronova E, Lobakova E (2014) Phycoremediation of alcohol distillery wastewater with a novel Chlorella sorokiniana strain cultivated in a photobioreactor monitored on-line via chlorophyll fluorescence. Algal Res 6:234–241

Somoza V (2005) Five years of research on health risks and benefits of Maillard reaction products: an update. Mol Nutr Food Res 49(7):663–672

Sowmeyan R, Swaminathan G (2008) RETRACTED: effluent treatment process in molasses-based distillery industries: a review. J Hazard Mater 152(2):453–462

Tamanna N, Mahmood N (2015) Food processing and maillard reaction products: effect on human health and nutrition. Int J Food Science

Taylor JL, Demyttenaere JC, Abbaspour Tehrani K, Olave CA, Regniers L, Verschaeve L, De Kimpe N (2004) Genotoxicity of melanoidin fractions derived from a standard glucose/glycine model. J Agri Food Chem 52(2):318–323

Thomsen M (2005) Complex media from processing of agricultural crops for microbial fermentation. Appl Microbiol Biotechnol 68(5):598–606

Tippayawong N, Thanompongchart P (2010) Biogas quality upgrade by simultaneous removal of CO_2 and H_2S in a packed column reactor. Energy 35(12):4531–4535

Tsagarakis KP (2007) Optimal number of energy generators for biogas utilization in wastewater treatment facility. Energy Conv Manage 48(10):2694–2698

Valderrama LT, Del Campo CM, Rodriguez CM, de-Bashan LE, Bashan Y (2002) Treatment of recalcitrant wastewater from ethanol and citric acid production using the microalga Chlorella vulgaris and the macrophyte Lemna minuscula. Water Res 36(17):4185–4192

van den Broek R, van den Burg T, van Wijk A, Turkenburg W (2000) Electricity generation from eucalyptus and bagasse by sugar mills in Nicaragua: a comparison with fuel oil electricity generation on the basis of costs, macro-economic impacts and environmental emissions. Biomass Bioenergy 19(5):311–335

D Vriens L, Franco P, Verachtert H (1985) 20th European brewery convention congress. Helsinki, pp 555–562

Wagh MP, Nemade PD (2015) Treatment of distillery spent wash by using Chemical Coagulation (CC) and Electro-coagulation [EC]. AJEP 3(5):159–163

Wagh MP, Nemade PD (2018) Biogas generation from distillery spent wash by using an OPUR western biotechnology process: a case study. Desalin Water Treat 118:241–248

Wang CY, Ng CC, Lin HT, Shyu YT (2011) Free radical-scavenging and tyrosinase-inhibiting activities of extracts from sorghum distillery residue. J Biosci Bioeng 111(5):554–556

Wolmarans B, De Villiers GH (2002) Start-up of a UASB effluent treatment plant on distillery wastewater. Water SA 28(1):63–68

World Health Organization (2018) COP24 special report: health and climate change

Yadav MP (2014) The role of biogas for environmental sustainability in Nepal: Users' Perspective. J Indian Res 2(3):49–56. ISSN: 2321–4155

Yentekakis IV, Papadam T, Goula G (2008) Electricity production from wastewater treatment via a novel biogas-SOFC aided process. Solid State Ion 179(27):1521–1525

Zhang Y (2008) Reviving the carbohydrate economy via multi-product lignocellulose biorefineries. J Ind Microbiol Biotechnol 35(5):367–375

Zishan F, Alayi R (2013) The role of biogas power plant in generating electric energy and its environmental effects. JMEAST 2(6):309–313

Chapter 9
Policy Implications for Sustainable Ethanol Production

Zeenat M. Ali, Syed Farman Ali Shah, Muhammad Mureed Tunio, Mazhar Abbas, and Muhammad Arshad

Abstract Economic, social, and political strength, development and activation depend upon the energy-related issues, demand, and the best solutions for the same. When the energy consumption patterns vary and the demand for the production increases the ways and means of energy do change with the pace of time and the fossil fuels were replaced by the diversified and focused on energy generation through nonrenewable sources. Bioethanol has been identified as the best alternate for the petroleum gasoline in the world, and it saves fossil fuels depletion and remedies of the greenhouse gases and employment opportunities. Residues of the food crops, grains, edible and nonedible seeds and vegetable oils may be the main feedstock for the biofuels in the world. Different feedstock with respect to the local and international trends of production of ethanol is mentioned and compared in this study. Impact of the production and the policy of bioethanol is worth discussed in the presentation. For the developing world, the food item being used for the production of bioethanol are discouraged and wastes or byproducts are preferred in this regard. Developing countries have shown their interest in the enhanced production of biofuels and many collaborative projects are in progress and policies, legislation is being formulated locally and international.

Keywords Bioethanol · Biofuel · Policy · Bioenergy

Z. M. Ali · S. F. A. Shah
Department of Chemical Engineering, Mehran University of Engineering and Technology, Jamshoro, Pakistan

M. M. Tunio
Department of Energy and Environment, Quaid-E-Awam University of Engineering Sciences and Technology, Nawabshah, Pakistan

M. Abbas · M. Arshad (✉)
Jhang-Campus, University of Veterinary and Animal Sciences Lahore, Lahore, Pakistan
e-mail: muhammad.arshad@uvas.edu.pk

M. Arshad (ed.), *Sustainable Ethanol and Climate Change*,
https://doi.org/10.1007/978-3-030-59280-6_9

9.1 Introduction

Energy is essential for all social-economic activates. The energy improves the standard of life and comfort. The rapid population growth, manmade activities, and urbanization increase the energy demands day by day. With pace of time, the energy consumption pattern changes. The emphases are on more energy production, energy sources, availability, losses, and end use efficiency (Ramchandra et al. 2000). Prior to the sources of energy were only the fossil fuels such as coal, coke, and petroleum products. Now the concept was diversified and focused on energy generation from nonrenewable sources.

The advancement of technologies and modernization enhanced the energy consumption in both industrial and domestic levels. The vehicles consumed the liquid petroleum fuels. The transport sector is a key component of social stability and economic growth (Nap 2009). The concerned is raised about the depletion of nonrenewable sources and future energy scarcity in Pakistan (Moriartv and Honnerv 2016).

The world now being explores the ethanol as best substitute of petroleum fuel. In regard Governments of various countries facilitate the production, promotion, and adoption of biofuels, especially ethanol and ethanol blend petroleum fuels. It is reported that forty countries already include ethanol in their energy matrix (Timilsina et al. 2011). Hence, the remarkable budget is allocated, given subsidy and exempted the tax exemption to promote the ethanol production in their counties (Sorda et al. 2010).

The main cause of switching to renewable sources is the depletion of fossil deposits, the remedies of greenhouses gases in climate, energy supply security and impending peaks and employment opportunities.

9.2 Sources of Ethanol Production in Pakistan

The biofuels are classified into first and second generation. The first-generation biofuels are produced from food crops, grains, edible and nonedible seeds and vegetable oils. The second generations are the residues such as corn stalks, cotton straw, wheat straw rice straw, leaves, and other ethanol bearing agrowastes. The ethanol produced from biosources is termed as bioethanol. The seasonal availability of various feeds stalks in Pakistan.

The corn gains are most popular for bioethanol production because of sugar, sucrose, and starch content. The other edible first-generation seeds are sunflower, rapeseed, mustard seed, and wheat grain. The nonedible seeds are jatropha, soap seed, and linseed. The main issue of first-generation biofuels production is that they are used as food or food stalks hence the production of bioethanol causes the shortage and scarcity (Naik et al. 2010). In this regard, worldwide various studies of different researchers are conducted and found that first-generation biofuels are not sustainable because it creates the potential stress on food commodities. However, the second-generation biofuels are more suitable and feasible (Sorda et al. 2010). As the second generations are the residue of agrowaste so they could possess the low procurement cost, high abundance, and current underutilization (Welfle et al. 2014). From energy security and environmental criteria, the production of ethanol from lingo cellulose is better than first-generation biofuels. The selection of feed stalk for bioethanol production strictly depends upon its composition including the cellulose, lignin, and ash content. The composition of various feeds stalks is given in Table 9.1.

According to the life cycle analysis, the ethanol production from lingo cellulose by biochemical reaction yields better. Further, the issue with lignocellulose is its tough and recalcitrant nature which increase the production cost. However, the cost should be reduced by enzymatic treatment and fermentation process. But prior pretreatment is mandatory (Bhutto et al. 2014). The number of researchers conducted research on production of ethanol from lignocellulose feedstock. Bhutto et al. (2014) reported that the process of ethanol production from feed stalks is more expensive. The researcher argued that the production steps increase the cost of operation. Another researcher addressed the issues and limitation raised during the process. These are the handling and transportation of biomass, efficient and effect process of separation of lignin from lingo cellulose agrowaste. These upraised the cost (Liew et al. 2014). It was reported that ethanol production from lingo cellulose stalks would not be feasible and suitable in every respect. Hence could not be recommended on commercial scale in near future (Carriquiry et al. 2010).

The source of ethanol production is molasses as well. The molasses is by-product of sugar industry. Pakistan is on world's 7th rank of sugarcane growing country

Table 9.1 Composition (dry basis) and maximum potential yields of different feeds talks (Bhutto et al. 2015)

Residue	Barley stalks	Cotton stalks	Corn stalk	Rice straw	Wheat straw	Sugarcane bagasse
Residue/Crop ratio	1.2	3	1	1.4	1.3	0.6
Crop dry matter (%)	88.7		86.2	88.6	89.1	26
Lignin (%)	13.8		8.2	5–24	8–15	18.4
Cellulose (%)	33.8	14.4	42.6	32–47	35–45	43.1
Hemicellulose (%)	21.9	14.4	221.3	19–27	20–30	31.1
Ash (%)	2.18	21.5	4.3	12.4	10.1	2.4
Biofuel yield (L/kg)	0.31	–	0.29	0.28	0.29	0.28
Biomass yield (k/ha)	1184	–	1734	1399	1413	11,188
Biofuel yield (L/ha)	367	–	503	392	410	3133

and sugar industrial sector is second-largest industrial sector of the country (Khan 2007). The ethanol is produced from molasses by fermentation process. The carbon content is very low so can generate the greenhouses gases in very low quantity (Dufey and Greig-Gran 2010). Ethanol synthesized from sugarcane emerged as a leading renewable fuel for transportation in USA (Araújo et al. 2017) It is also blended with gasoline and can reduce petroleum usage (Dias De Oliveira et al. 2005). Approximately 80% of cane molasses is used for production of alcohol in the world.

9.3 Ethanol Prospectus in Pakistan

In Pakistan, 48.7% petroleum products were consumed in transport sector in the year 2013–2014. Government of Pakistan imports the petroleum fuels of cost $4 \cdot 310^9$ USD in 2014 in order to meet the country's demand (Karim and Shahid 2018). The oil reservoirs deposits are depleting day by day. The country is running on deficit budget. Hence, this huge amount puts the financial burden to economy. The bioethanol production profile is shown in Fig. 9.2

Today's need is to explore, to develop the indigenous sources of ethanol production so to resolve the problem of energy crisis and to establish the well-develop technologies on commercial scale. It also matters to fix the stable ethanol prices in local market. The Pakistan State Oil (PSO) imported the blend contents 10% anhydrous ethanol and 10% gasoline in 2010 and started its marketing in Karachi (Bhutto et al. 2015). The price was fixed by Pakistan Gasoline Regulating Authority (OGRA)

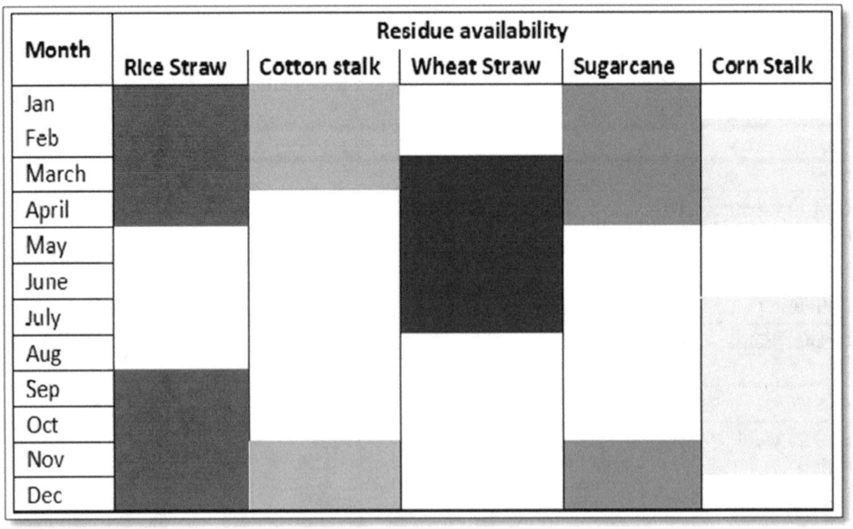

Month	Residue availability				
	Rice Straw	Cotton stalk	Wheat Straw	Sugarcane	Corn Stalk
Jan					
Feb					
March					
April					
May					
June					
July					
Aug					
Sep					
Oct					
Nov					
Dec					

Fig. 9.1 Seasonal availability various feed stalks in Pakistan (Bhutto et al. 2015)

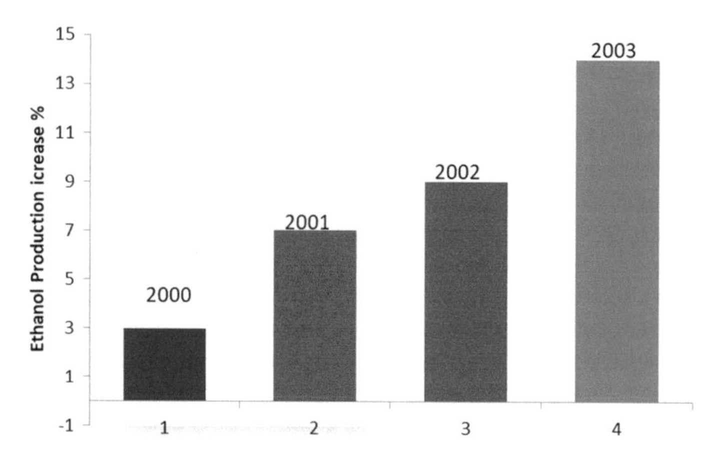

Fig. 9.2 Bioethanol production (%) in Pakistan (Dufey and Greig-Gran 2010)

4% less than regular gasoline to facilitate the public. This offer exhibited positive impacts now the demands of ethanol in the country became $364 \cdot 10^6$. Pakistan Sugar Mills Association (PSMA) is taking keen interest in development of the bioethanol advancement program. The Ministry of Food Agriculture and Livestock (MINFAL) has also urged to find out other sources of raw material for bioethanol production such as wheat, potatoes, rice, and sorghum (Asif 2009). Production of bioethanol from cane sugar is economically friendly than other sources (Canilha et al. 2012).

Presently molasses is placed the prominent position in ethanol production. The quality of molasses matters roughly 1-ton molasses produced 240–270 L ethanol. The production of 500 million tons ethanol per annum contributes to 36% gasoline consumption in transportation sector of the country (Harijan et al. 2009; Harijan 2008, PhD thesis).

Asif (2009) reported that there were thirteen distilleries for ethanol production. The three distilleries were not attached with sugar industries having annual capacity of 6.5 million liters. The remaining distilleries are attached but two of them are not working. The annual ethanol production rate was about 143 million liters verses the molasses consumption 0.560 million tons. The current production of ethanol is 274 million annuals (Asif 2009). A report of 2007 showed that only 6 sugar mills had a facility of processing of molasses into ethanol from the total sugar mills. With the current production level of sugarcane crop, Pakistan has a potential to produce over 400,000 tons of ethanol (Balat 2009).

9.4 Impact of Bioethanol Production

It is the fact that last five years the biofuel consumption rapidly increases and its growth into a global industry. Many rich, low, and middle countries planned to enhance the production, strength the technology and improve the process. Globally the policies and ambitious target planned to promote the significant growth. In 2007, globally 1% biofuel was reserved for transport sector, however, the estimated biofuel production was 54 billion liters (Dufey and Grieg-Gran 2010).

The main aim of biofuel production especially industrially developed countries was to mitigate the greenhouse gas. On the other side, in the less developed low and middle countries the driven force is to get rid from energy scarcity, earn foreign exchange by export the biofuel and provide the employment to local population. Also, to support the rural development programs. It provides the greater energy security.

The liquid biofuel can be produced from wide range of agricultural commodities. The controversial opinions are developed by various organizations. In the report 2008 of Food and Agricultural Organization (FAO), it was reported that next decade and beyond the biofuels will be produced from agricultural commodities which put the stress on consumer market. The stakeholders, local growers, public, and NGOs have raised their concerns in this regard.

Besides these, the other issues have also been associated with production of ethanol on large scale. These are the intensive agricultural puts the adverse effects on environment. The soil erosion would be possible. The high use of water causes the water scarcity in the region. The uses of agrochemicals can disturb the ecosystem. The biodiversity would happen. The agricultural run-off pollutes the water bodies. The use of palm oil and soy causes deforestation in Asia and Latin America.

In contrast, another study urged that biofuel is beneficial to climate change. It is the mitigation to climate change. The biofuel accounts the processing and cultivation process also taken part to cope up the energy shortage.

The extensive use of biofuel production matters the rise up in food prices and puts impacts on food security. The studies examined that the food prices impact will be 3–75%. With median range 30–40% for a maze and somewhat less for other basic commodities (FAO 2009).

The commercial cultivation of food commodities for bioethanol production would cause the displacement poor people who depend on land. This is a social issue observed in many parts of the country.

9.5 Bioethanol Trends

The consumption of biofuels increases day by day. The developed world such as United States, Brazil, and member countries of European Union put the hands together to make the biofuel program successful. The faxable polices, agricultural reforms, and tax reduction/exemption facilitate the bioethanol producers. Due to Government intervention, globally the bioethanol demand will increase by 125 billion liters in 2020 (Demirbas and Karslioglu 2007).

The Global production of bioethanol increase from 2000 to 2007 was 17.25 billion liters (Balat 2007a, b) to 46 billion liters, respectively. The global consumption of gasoline was 13 billion liters in 2007 from that the 4% consumption was of bioethanol. The United State is the largest bioethanol producer. It contributes globally

47% of total bioethanol production. The total production in 2005 was 15 billion liters (Renewable Energy Network for the 21st Century (REN21) 2008) and in 2006 it reaches to 18.3 billion liter (Renewable Energy Network for the 21st Century (REN21) 2007). The Energy Policy Act 2012 (EP Act) target was 57 billion liters (Balat and Balat 2009).

Brazil is the second largest bioethanol producer country and topmost bioethanol exporter. Most of the fuel used domestically is only 20% exported to European Union, United State, and other markets (https://agb.east.asu.edu/workingpapers/0207.pdf). In Brazil, most of bioethanol produced is from sugarcane industry and fulfills the 40% Brazil petroleum consumption (https://www.greenergy.com). Between the Brazil and United states, the Memorandum of Understanding (MoU) was signed in March 9, 2007. Both countries agreed in mutual cooperation and coordination. The advancement in research will be shared. Both countries pledged to switch their industries to biofuels and efforts the worldwide development of biofuel (Seelke and Yacobucci 2007).

9.6 Policy Drivers for Bioethanol

Brazil is the topmost bioethanol producing country. The history starts up since long. Currently, the country is self-sufficient and highly capable in technology. The workers are skilled and fully potent in bioethanol production.

This happened in 1975 when the national alcohol program (ProAlcool) was initiated in Brazil. The aim and objective of that program was to reduce or minimize the use of gasoline and to discover the best alternate of the gasoline. As the gasoline is deplete and scare. The prices rise up to sky and are emitting the greenhouse gases. Hence, the attention was paid to biofuels especially bioethanol. The Government took interest and made the effective steps in this regard. The Government legalizes, mandates the legislation, design policies, and plans, provides facility and finance, encourages the venders and bioethanol producers, creates the skill development opportunities. The Government gives the subsidy and tax exemption to bioethanol producers. These Government efforts were creating new spirits and bioethanol industry to move toward success. The demand of bioethanol was going to increase and the industry gradually scale up. The more advancement in this field was incorporated.

Till 1984, the new cars fueled by anhydrous ethanol were introduced in the market. In anhydrous ethanol content, 96% is ethanol, the rest would be the water (Kline et al. 2008). In 1993, the Government law was passed according to that the venders were bound to sell the blended ethanol 20–25% in market (Martines-Filho et al. 2006). In 1999, The Pro Alcohol program was on way to success and some changes were made as the public participation was increased their rights were reserved and offered more incentives. The Government interventions were minimized. The public was authorized for price allocation verses profit margin. By the support of Government, the flexible-fuel cars were introduced in markets. The rising prices of gasoline pushed toward the utilization of bioethanol. The other options were also in queue including the bioethanol production from sugar industry. That was why the sugarcane production and sugar industry strengthened in 2000 (Kline et al. 2008). Currently, the Brazil automobile industry manufactured 80% flexible fuel cars. Among that the 30% was done in 2004. As the bioethanol is environment friendly and viable hence 32,000 stations provide the services to public. The public avails the facility according to their will and wish. They have the option of anhydrous bioethanol/gasoline and 25% bioethanol/gasoline (Coyle 2007).

The European Commission (EC) is promoting the bioethanol industry in their member countries. The scope was to mitigate the greenhouses gases emitted from gasoline and transport sector is topmost contributor. Secondly to gripe the issue of energy scarcity and improve the energy security. Thirdly to develop the rural areas by means of employment, health, and education (Balat 2007a, b; Jansen 2003). It emphases to reduce or get rid of gasoline fuel and minimize the dependency on it. The transport sector fuel consumption accounts for 30% of total energy and it goes high. This is because of mobility and good transportation. According to European Union white paper (https://www.europa.eu.int), there should be the need to implement and enforce the viable plans and practice the sound strategy to switch to the alternative fuels (Malça & Freire, 2006). In that regard the directives (2003/30/EU) set by European Union that in transport sector 5.76% biofuel were consumed (European Commission (EC). Directive 2003). The member countries unanimously agreed and included in their national objectives. Some countries were ambitious and set their target such as France (Wiesenthal et al. 2009) and Belgium (European Commission (EC). Green Paper 2006) set its target as 7% in 2010. In 2015, the target was 10% by France (Wiesenthal et al. 2009).

The various strategies that were in records more focused on alternative fuel and reduction in greenhouse gases. The green papers "A European Strategy for Sustainable, Competitive and Secure Energy" Published in March 2006 by the European commission (European Commission (EC). Green Paper 2006). And the next year that was January 2007 the energy review report "An Energy Policy for Europe" was cited (European Commission (EC), Jan 2007, Strategic energy review). It is said that to make the environment more friendly, open the new doors for alternative fuels and promote in markets to ensure the progress, prosperity, and growth in all sectors of life (European Commission (EC) 2008). The purpose was to achieve the target of 10% renewable energy set by European Union extended to 20% in 2020. Also issued the fuel quality directives in 2010 beneficent to save up to 35% greenhouse gas emissions (Pfuderer and Castillo 2008; Thamsiriroj and Murphy 2009). By the tax exemptions, lenient financial policies the bioethanol industries grown-up (Jacquet et al. 2007). The magnitude of excise tax levied on petroleum products was high compared to bioethanol. This made the better consumer choice of bioethanol.

The fiscal instrument depends upon the reduction in excise tax on biofuel and makes the bioethanol consumption viable. This required high tax exemption on bioethanol and less tax reduction in petroleum products (Kojima et al. 2007).

The European Union directs its member countries about the taxation framework. It allows the tax reduction in the favor of electricity and renewable energy. European

Union said that loss in tax revenue generation were covered from biofuel intended for transportation and it may not less be than 50% of normal excise duty (Schnepf 2006). In European Union countries, the tax reduction is as high as US$ 0.84 per lit (Kojima et al. 2007).

The Chinese run their renewable fuel program by making amendments in law. They made the legislation regarding the tax levied, tax exemption or reduction, price limits, and subsidies (Wang et al. 2006). In 2001, the Chinese Government initiated two programs about renewable energy with aim to improve the rural and domestic environment and bioenergy generation (Gnansounou et al. 2005). The more improvements were incorporated in law and promulgated. These strengthen the bioethanol industry and make it self-sufficient (US International Trade Commission July 2008). In Feb 2005, the renewable energy act was passed according to that the 10% energy will be biofuel since 2010 (World watch Institute. State of the World 2006). The government set the selling price of bioethanol US$ 0.84 per lit. This is 91.1% selling price of gasoline (Yang and Lu 2007).

The Pakistan Government pledged to switch the alternate source in order to overcome the shortage of energy within the country. The Hydrocarbon Development Institute of Pakistan (HDIP) is working on that grounds and conducting the feasibility survey. In support, the Pakistan State Oil (PSO) lounged the pilot plant outlets in major cities of Pakistan such as Islamabad, Karachi, and Lahore to monitor the 25 preidentified vehicles for six-month period. The outlet provides the 10% blended bioethanol fuel (https://www.dawn.com/2006/07/28/nat1.htmS).

The United States (US) gave the bioethanol production at its priority list particularly from maize. They reform in agricultural sector since 1980 and revitalize their agricultural policies in farming sectors (Jull et al. 2007). The new vehicles are marketed which performed on blended fuel (E85). The E85 contents 15 and 85% gasoline and bioethanol, respectively. This consumed 1% of total bioethanol produced in United States (Yacobucci and Schnepf 2007).

The US congress promulgated the use of the bioethanol and other alternatives fuels. The legislations and amendments were passed in terms of incentives and promotion. The significant step was the enforcement of The Energy Policy Act 2005 (EPAct) (Hoekman 2009). In that connection, the legislation The Renewable Fuel Standards (RFS) was set. For 2012 targeted the consumption of 5% gasoline verses the 28.4 billion bioethanol (Jank et al. 2007). In 2008, this act offered the more incentive to the production of bioethanol and biodiesel from the cellulosic materials. The small producers of biodiesel were levied the income tax US$0.03 per liter (Kojima et al. 2007). The gasoline consumption became high in 2007 and records 9.7 billion per day In December 18, 2007, the US Congress passed the legislation Energy Independence and Security Act (EISA) undersigned by President of United State unanimously agreed to improve the fuel economy and reduce the dependence on foreign fuels. This act also represents the standards of carbon fuel cycle and notify the 20% carbon intensity reduction (Kojima et al. 2007). The Cooperate Fuel Economy Standards (CAFE) were introduced. According to that by 2020 the light trucks and car the fleet-wide average of 35 miles per gallon. This act also notifies that in 2008 the 34 billion liters biofuels specially ethanol and in 2012 the 57.5 billion liters in 2020 the 136 billion liters would be required.

9.6.1 Socio-economic Impact of Bioethanol Production

The social-economic impacts of ethanol production industry refer to these issues relevant to establishment of bioethanol industry and their consequences on individuals and as well whole.

9.6.2 Live Hood Services

It plays the vital role in development, progress, and prosperity of people. It puts the long-lasting socio-economic impacts. The sustainable and eco-friendly way would improve the standard of life and give comfort to life while nonsustainability will charge the price and one have to pay. This may be long term or short term and depends on process followed.

The bioethanol plant prominently puts the direct and indirect effects on rural development. It provides the job opportunities to local resident. The employment will improve the skills and play role in capacity building. The employs would get incentives and salary which makes their life happy, satisfied and improves the life standards such as diet, health, and education. The new hospitals equipped with latest machinery and advanced diagnostic centers will be possible to establish. The drug storage facilities, sterilization, and pathological labs will be improved.

By the production of bioethanol, the governments will overcome the issue of energy crises. Besides all direct benefits, there are indirect benefits as well. Those include the establishment of new industries which consume directly the bioethanol in their products also those industries who consume the end product and residue of bioethanol industry. This scenario implies that new job vacancies would be generated. In 2017, the unemployment rate in Pakistan was 5.9% by this way the unemployment rate would be reduced. More people would engage in various jobs. The doors would be open to new organizations who involve in marketing of bioethanol. The bioethanol production will strengthen the other sectors of national development like transport, communications, and infrastructure in educational, health, and other sectors.

This is general practice that people displace from low developed areas to high. The establishment of bioethanol industry would mitigate the migration from rural to urban areas that make the urban load balanced.

The bioethanol production will boost the economy and provide the financial support. By this way, the things would be improved toward the betterment. When economy improves the purchasing power will also increase. Hence the buy and purchase activities increase. This generates the revenue and government will receive in terms of taxes.

The consumption and demand of local goods will increase. The people will hire the local services which support and strengthen the local industry. And their positive impacts would convey to local people.

The extension of bioethanol industry compelled the local farmer for more production to cultivate the diversified seeds. This can generate the competitive environment and increase the profit margin. This causes more foreign exchange when commodity possessed the high cost in foreign market. The highly satisfied and financially stable life leads to high purchasing power. The selling and buying will generate more revenue and government gets more taxes. The agrochemical products such as pesticides, fumigates, manures, plant enhancers, insecticides, and fertilizers industries would be grown-up. Further, the agrotool's demand will be increased.

9.7 Biofuel Policies in Major Biofuel Producing Countries

The great foresight, vision and professional experience are mandatory to design viable biofuel policy for the long interest of the countries. It should be compactable and sustainable in long run. Secondly, it should be reviewed, revised, and amended subject to the condition in order to meet the goal. The point of view is to switch to renewable source, to save the nonrenewable source, to reduce their depletion, to minimize the greenhouse gas emission, and develop the stability between food and fuel chain. Here, we are going to discuss the policies of various countries (Chun Sheng Goh 2010).

9.7.1 Malaysia

Malaysia started its journey in 1979 by announcement of its National Energy policy, with clear objectives of supply, utilization, and environment (Hitam 1999). In 1980, the Depletion Policy was run to conserve the energy sources. In 1981, Fuel Diversification Policy was implemented to avoid the dependency on single energy source. In broad spectrum, four fuels were incorporated in energy matrix. The renewable sources were kept on priority. The dependency and consumption of oil, coal, and natural gas were reduced from 80% to less than 10% during 1980–2003. The substantive issue was the excess emission of greenhouse gas that was 237% more than the permissible limit. As the part of UN Convention on Climate Change and the Kyoto Protocol, it was mandatory to reduce the greenhouse gases. In this regard, the Malaysian Government more focused and induced the renewable source as fifth in energy matrix. In 1999, the 8th Malaysia Plan (2001–2005) was on table in which it was emphasized to generate the 5% energy from renewable sources. This would save the amount of RM 5 billion (Leo-Moggie 1996).

As the matter of fact, the renewable energy infrastructure in Malaysia was under developing stage when it runs full at its optimum capacity, it would minimize the fossil fuel dependency and would cause the clean environment. This planning was strengthened by accomplishment of Small Renewable Energy Power Program in 2001. According to that the Electricity was generated by renewable source and supply the same to Electricity generation Unit to enhance its generation capacity up to 352 M. The renewable energy sources mean solar, geothermal, tidal, wind, and biomass (Ministry of Energy, Telecommunications and Multimedia) 2004]. It was worthful that among all renewable sources the biomass was most feasible in a manner that huge amount of agricultural waste was generated annually. The utilization of agriculture waste would be safe and economical. Under the umbrella of Government of Malaysia, the biomass power generation and cogeneration project were run in October 2002. This not only reduces the greenhouse gas content in environment but also effective utilization of palm oil waste cogeneration (PTM (Malaysia Energy Centre) 2004). Through cogeneration second-generation stalks are utilized as palm oil residue and other agricultural wastes. This would make more applicable and practicable by proper planning, specific strategy, technical expertise, and legislation (Mohamed and Lee 2006).

It is noteworthy to mention here the example of Brazil. In 1975, Brazil launched the Brazilian Alcohol Program (ProAlcohol) in connection to reduce the 80% fossil fuel consumption (Tan et al. 2008). The concurrently scenario indicated that surplus bioethanol production come up in Brazil that was 164 million m^3 verse to consumption of 141 million m^3 gasoline (Goldemberg et al. 2004). Malaysia and other countries would have to adopt such a wide policy that they would be self-sustained countries. Toward the strengthening, it is suggested to offer attractive package to Bioethanol Producers as subsidy and low markup loans/interest-free loans. Also, award the funding on research and development projects. Encourage the growers of second-generation lignocellulose and provide them storage and transportation facility. It is mandatory to provide technical assistance and develop the coordination plan.

Besides all the firm planning and efforts, the Government of Malaysia did not achieve the goal of 5% bioethanol blending. The National Biofuel Policy stated that failure lies because of disorganized and disordered policy played by Ministry of Plantation Industries and Commodities, 2006. The main reason was unstable and fluctuated prize of volatile palm oil in national and international markets. The insecurity of raw material supply created the uncertainty in economic Policy. The streamline and definite availability of raw material for bioethanol industry is very important to reduce risk factor for investors.

In Malaysia, the 91 companies were registered but only 2 were involved in bioethanol industry (Nagarajan 2008). Their interest and attraction in bioethanol production are not looking so high. On the contrary, the automobile industries did not letdown their dependency on coal and petroleum products. They turned to other cost effective and economical source as hydroelectricity and coal power plants (PTM 2008). The implementation of renewable energy matrix in country was not succeeded. The need was The Kyoto Protocol should be promoted as well as pragmatic planning should be followed (UNFCC 2008).

The bioethanol is hydroscopic in nature and has lower boiling point fuel and enables the corrosion in engines. Hence, it should be consumed in blended form. Because of inherent property to hold water molecule it creates difficulty to start the engine. The engine should be modified prior to use the bioethanol. The lower boiling point causes the vapor lock during the hot weather. The special measure is needed during transportation and storage. Only the flexible fuel vehicle permits the high percentage of bioethanol in blend.

The flexible fuel vehicles are most popular in Brazil the 70% of total sale volume while 3% in USA (Tan et al., 2008). This happens by mutual guardianship of Government and public acceptance and support.

The Proton Berhad and Perodua Sdn. Bhd big car manufacturing companies of Malaysia committed to manufactured the flexible fuel vehicles in future with aim to provide the luxury to consumer and to switch on bioethanol.

9.7.2 Brazil

Brazil is famous for bioethanol production. The bioethanol demand in 2012 was 83.1 billion liters holistically among that the 3% consumed by transportation sector. Brazil produced 1% and remaining by Europe, China from corn and Latin America (REN21 2013).

Brazil cultivated 51% sugarcane to meet the demand (Brazil 2012a) and planned to increase its production from 21 to 61.6% billion liters till 2020 (Brazil 2012b). The 27 countries of world were aimed to switch to renewable energy sources but only Brazil and USA would firmly implement the policy and acquire the result. It was assumed that bioethanol was produced from second generation so as to reduce load of food chain but the matter was there was no full-fledged technology still developed to convert the lignocellulose material to bioethanol on economical basis.

Brazil preferred to produce from sugarcane because large cultivated land was available. It was aimed to increase the capacity of ethanol the USA, Brazil, and other countries were planned to enhance their production 79.8, 24.2, and 138.8% up to 2021 categorically. In 2015, the existed capacity of these countries was 57.1%, 37.4%, and 138.8%, respectively.

The point of ponder was well-justified land used policy for cultivation of energy crops (HLPE 2013). No doubt the world population rapidly increases simultaneously the food, water, and energy required for life. In this regard, the sustainable policy was formulated for security of food chain and energy crops (RES-D 2009).

Brazil took solid steps toward the sustainable path and stood at prominent position among hundred countries (FAO 2013). The global economy was adopting the sustainable and renewable roadmaps (Goldemberg 2007). This was because the full dependency of fossil fuels was not feasible for them. The fossil fuel reservoirs are depleting day by day (Kerr 2011). Possibility that their production will decline in coming decades (Hallock et al. 2004, 2014). Other than fossil fuel dependency were increase the choice of customer and increase the security of energy (Chu and Majumdar

2012). Also eliminates the greenhouse gas emission release during combustion of fossil fuels. Besides all the improve the better social economic environment, creates the job opportunities, provide employment and open the new horizons. The foremost contribution would be the favorable support to our ecoservices as food, water, and others. This happens when comprehensive and widespread practices are adopted, do not deforest as it leads to biodiversity (Cerri et al. 2009).

The Government of Brazil implemented the National Plan on Climate Change in 2019 to reduce the greenhouse gases emission. According to this, the farmers were encouraged to adopt the best agricultural practices. The Local Control and Accountability Plan were introduced to offer US$ 90 billion loan to growers in account to enhance their productivity till 2020. It was aimed to develop the grassland and pastures of degraded land. It was noted that 7.6% land in 2011 was barded and undeveloped. This was included in planning that with integration of livestock the 4 Mha land was cultivated.

The reason behind that was to improve the soil quality by addition of animal manure to give rise the nitrogen content. This increases the crop yields by 11% and pasture supporting capacity by 1.76% prior it was 1.05 (Landers 2007). This was promoting the reforestation and more land available for agriculture.

9.7.3 European Union

In 1983, in Italian Ferruzzi group emphases the Production of biofuel from agricultural residue, grain, straw, and wine waste and to exterminate the lead content fuel (Londo and Deurwaarder 2007). Generally, the lead is added in the fuel to enhance the octane number of fuel and to generate the anti-knocking properties of fuel.

The European Commission started its beginning innings to discover the scope and opportunities of biofuels in the region. In 1990, the efforts were more focused. In the same year, Austria developed the first commercial Biofuel Plant (Körbitz et al. 2003). Sweden took step in 1991 with the support of rural Sweden farmers in order to secure the availability of raw material for biofuel industry (Grahn 2004). In Germany, the biofuel blended with mineral oil was completely tax exempted. In British, the tax was not exempted as it was the contravention of European Energy Tax Directive Contravention.

The European Commission in 1997 published the White paper about Renewable Energy and termed as Energy of Future. The tax was exempted 2% in blended biofuels for transportation sector. Resulting biofuel consumption increase in 2010 was 18 Mtoe (Thuijl and Deurwaarde 2006). It was the great scope of biofuel in Energy sector and would place in prominent position in energy matrix. It should be considered as best substitute for fossil fuel. By induction of biofuel, it managed the consistent supply of Energy to consumer. Energy shortfall should be covered specially in Developing Projects.

In the year 2000, the Organization of the Petroleum Exporting Countries puts restriction in oil exports so the oil prices raised to $30 per barrel. At that time the

European Commission explores the alternative fuel in order to ensure the security and continuous supply of fuel. For the year of 2020, it was targeted to consume 20% bioethanol of the total fuel (European Commission 2000). The tax was exempted and facilitates the vendors and producers of biofuel.

The transportation sector is major consumer of fossil fuel hence in 2001 Policy of Biofuel was formulated in the way to unfold the scope of alternative fuels so as to ensure the security and supply of fuel and minimize the dependency. It was planned that alternative fuel contribution was up to 2% and with increment rate of 0.75% in five years from 2005 to 2010 (European Commission 2001).

To enhance the production, the tax was exempted and encourages the member countries even those who provide the assistance to developing countries should also be exempted. The execution of such practice was not fully supported and liked by France and Germany as there were the largest biofuel generating countries. The other member countries such as Spain, Italy, and Austria set their target and planned strategies but finally it was realized that the capability and potential vary hence the countries should plan and implement by themselves. Keeping in view their resources, potential, manpower and technical expertise (Aantjes 2007).

On the platform of Kyoto framework, the Energy security supply was again ensured. The relevant issue such as supply of raw material, technology frame work, effects on climate, imports of oils, and economic issues were addressed in 2003. It was also kept on board the development issue of rural areas, changes in climatic and emission of greenhouse gases. It was notifying that bioethanol industry should meet the standard from production to end user (European Parliament, 2003).

It was worthful to declare here that only Germany and Sweden achieved the target of 2% set by European commission while the other member countries were stood on 1%. The plan was not fully succeeded. The reason behind was the variant environmental conditions and social circumstance. Hence, it was stated that instead of following the directives of European Union the individual countries should plan according to their own expertise. The European Union again presented "Biomass Action Plan" for large-scale biofuel production, consumption, and measures (European Commission 2005).

9.7.4 Thailand

The biofuels abundantly consumed in transport sector. Its consumption increases day by day. This increment was recorded up to 23% from 2001 to 2007 (Yan and Lin, 2009). The various concerns were related to biofuel productions on commercial scales. These were land use, effect of soil quality, and security of food crops. It was a general observation that if farmers and growers get more incentives, highest crop rate, encouragement, and friendly polices from government and corporate sector of bioethanol producers then there is possibility of scarcity, inadequate, and insufficient food crop supply observed in long term. In that context, the poor countries would face more consequences and the food price increases. The need was to execute the policy which explores the viable solution (Ewing and Msangi 2009). The excess cultivation of crop causes the soil infertility and nonarable (WBGU 2010; UNEP 2008). It was assumed that agricultural waste and inedible crops were taken into account for biofuel production. The aim of Government of Thailand in 2004 was to reformation and strengthen the rural area and lower down the oil imports. At the beginning, the Cassava and molasses were under process. The ethanol blends in Thai market were E10, E20, and E85. The E10 and E 20 mean 10% and 20% of ethanol blended in Petroleum introduced in 2004 and 2008, respectively. After August 2008, the E85 was induced in Thai Market. It is fact that bioethanol market from 2006 to 2009 was from 0.37 M liter/day to 1.03 M liter/(DEDE 2009a). It was lined up in 15 years plan from 2008 to 2020 that attention should be paid on bioethanol production. The short term up to 2011, midterm up to 2016, and long term 2022 policy were formulated. It was planned that up to 2011, 2016, and 2021 the 3 M, 6.2, and 9 L/day were achieved, respectively. Thailand is major food exporter country. The country totals 51.31 million hectares. In 2007, the 20.85 million hectares are reserved for agricultural. This stood 40% of total land (OAE 2008). The rice is major crop covers 48%, fruit 22% and sugarcane, Cassava and other fruits 21% of total of agricultural crops. For bioethanol production, the Cassava and sugarcane were considered for commercial production because of their surplus harvesting in country. The Thailand official data (2008) of 46 sugar mills reveals that 46 kg molasses, 104 kg sugar were generated from the processing of one-ton sugarcane (OCSB 2009). The 78% molasses was locally consumed among which the 37% to Bioethanol Industries and 30% to distilleries. The surplus was exported (DEDE 2009b). The 12.5–14.3 kg sugarcane generate per liter ethanol (Macedo et al. 2008). After Brazil, Thailand is one of topmost sugarcane producing country (OAE 2008). In Thailand, the sugarcane cultivation harvesting period is 10–12 months. The harvesting time is 3–4 months starts from December to March. There are three cultivation region northern, northeastern, and Central region yields 37%, 38%, and 35%, respectively. All these three regions exist on tropical and subtropical region.

The Cassava is used for bioethanol production. It is also considered as energy crop. It exists in sweet and bitter taste. This is because of presence of hydrocyanic acid. The sweet Cassava is edible and has low content percentage of hydrocyanic acid. While the bitter is vise versa. The sweet Cassava is not commercially cultivated and have no demand in local market. It is cultivated in house and small farm houses (TTDI 2004). The Cassava is the most suitable energy plant that contains high-quality carbohydrate, has high resistant against climate and low cost of cultivation. It is important that around the year in all seasons it can be cultivated. The nonedible Cassava is most feasible and the starch content was converted by acid treatment/fermentation/eczematic treatment into sugar. It is used as sweetener (Sriroth and Piyachomkwan 2008).

The Cassava fresh roots and in dried form is used for ethanol production. The dried is more viable. The starch content in Cassava roots was 25%, through fermentation and saccharification converts into one-liter ethanol. This is most feasible for commercial-scale production (Sriroth and Piyachomkwan 2008). Thailand known by the agricultural crop production including the energy crops such as maize, Cassava, sugarcane, and sweet Sorghum. Among all the sugarcane and Cassava are most feasible. The Cassava is highly cultivated in Thailand and captured the 70% of total market. Though the Sweet Sorghum possessed high potential content the Saccharose and Fructose, its cultivation trend is low. On positive aspect of Thailand is that the country has high scope to generate the bioethanol from second-generation crops and agricultural waste such as rice husk, wheat straw, corn stalk, cane trash, and palm oil fruit bunches and trunks. It is expected that in 2022 the total capacity of bioethanol production would lie between 3.6 and 17.6 M liter ethanol/day relevant to energy crop yield. The Cassava and sugarcane are the most potential crops for bioethanol production. In 2008, the 92% bioethanol were produced from sugarcane molasses remaining from Cassava. In the country, the Cassava is the most abundantly grown-up about 3.5% surplus. In 2016, 24 Cassava-bioethanol plants were in operational. In 2016 and 2022, the shortfall of Cassava was observed at about 6.95 and 20.63 M ton compensated by decrease in export by 24–55%. There are 19 bioethanol plants out of 48 who have capacity of 3.43 M liter ethanol/day run on either on sugarcane/Cassava or both simultaneously. The complete dependency on sugarcane is not feasible on long run, however, gives the price stability to sugarcane growers. The bioethanol from sugarcane would give more opportunity as sugar compared to Cassava. The concrete economic policy and market mechanism to ensure the prices of bioethanol, sugar, and all other supplementary items (Kaltner et al. 2005). The social acceptability of bioethanol is high. The results showed that bioethanol produced from Cassava and molasses reduced the greenhouse gases emission by 64% and 49%, respectively. It bit varies depend upon various factors including the operating condition, processing parameters, methodology, and conversion factor (Silalertruksa et al. 2009).

The agriculture residue or second generation most attractive offer because no need of any extra land for cultivation of energy crop. The bulk residue is generated in fields and often not utilized usefully in value-added products. The conversion in biofuel tends to safe disposal of agricultural waste. It is important that it does not put any burden on food chain. The food chain is either intercept or interrupt in long run. Could improve the diversification of feedstocks. The residue availability is 28.2, 21.9, and 9.6 M ton/year. It was perceived that oil palm fronds, sugarcane trash, leaves, and tops left in agricultural fields can be burnt to prepare the land for new crops. The burning smoke is an environmental concern (Robertson 2003).

It was endorsed in 15 years plan from 2008 to 2022 that major feed stalks for bioethanol production would be molasses, sugarcane, and Cassava. The other aspects were the cycles of greenhouse gas emission and feed stalk supply. The surplus availabilities and the net feedstock balances indicate that the total capacity of bioethanol production in Thailand in the year 2022 could vary from 3.6 to 17.6 M liter ethanol/day. It was documented that life cycle of greenhouse gas emission was 64%, 49%, and 87% when bioethanol was produced from molasses, Cassava, sugarcane juice, respectively.

9.8 Conclusion

It is concluded that bioethanol plays a prominent role in energy sector. The fossil fuel is depleting day by day. The consumption is increasing because of rapid population growth and technology boom. The only dependency on fossil fuel is not applicable in future. The induction of bioethanol in energy market would overcome the shortfall of energy.

There are various energy crops for bioethanol production including the sugarcane molasses, Cassava, and others. The bioethanol from sugarcane has more opportunities to produce sugar and other products. The matter is to formulate the long-term policy to ensure the sugar price in national and international markets.

It was recommended that second generation is more feasible for bioethanol generation. Specially the agricultural waste and residue. For stable long run, bioethanol production is possible when the feedback supply is streamlined. The bioethanol production plant should have dual and multiple option to deal with agricultural waste and lignocellulosic material at a time or switch to either one.

The bioethanol is hydroscopic in nature, causes corrosion because it holds the inherent water molecule. That is why it is not viable to use separately. The blend with petroleum is feasible. In different countries, it is available with various brand names.

The blend bioethanol can be used in petroleum-fueled vehicles and need no alternation, however, the lower boiling point causes the vapor lock during hot summer season and cannot start the engine promptly. It was a suitable option to vehicle manufacturer to design the flexible fuel vehicle in connection to consume the blended bioethanol.

The petroleum products rise up the greenhouse gases in environment. The use of bioethanol minimizes the level and creates the environment clean.

The concern raised by environmentalist that excess cultivation of energy crop would damage the agricultural land. This issue is resolved by specifying the area for particular crop and the establishment of livestock forms. The animal dung would be utilized as natural manure and enhanced the nitrogen level of soil.

The bioethanol production plant would accelerate the progress wheel. The GDP of countries would boost. The new employment opportunity increases. The rural areas go toward progress and prosperity.

References

Aantjes JC (2007) Driving biofuels in Europe: a research on the interaction between external regulation and value chain governance. Master Thesis. Rotterdam: Erasmus University

Araújo K, Mahajan D, Kerr R, Silva MD (2017) Global biofuels at the crossroads: an overview of technical, policy, and investment complexities in the sustainability of biofuel development. Agriculture 7:32

Asif M (2009) Sustainable energy options for Pakistan. Renew Sustain Energy Rev 13:903–909

Balat M (2007) An overview of biofuels and policies in the European Union countries. Energy Sources Part B 2, 167–81

Balat M (2007) Global bio-fuel processing and production trends. Energy Explore Exploit 25:195–218

Balat M (2009) Bioethanol as a vehicular fuel: a critical review. Energy Sources Part A 31:1242–1255

Bhutto AW, Bazmi AA, Zahedi G (2011) Greener energy: issues and challenges for Pakistan biomass energy prospective. Renew Sustain Energy Rev 15, 3207–3219

Bhutto AW, Harijan K, Qureshi K, Bazmi AA, Bahadori A (2015) Perspectives for the production of ethanol from lignocellulosic feedstock–a case study. J Clean Prod 95:184–193

Bhutto AW, Qureshi K, Harijan K, Zahedi G, Bahadori A (2014) Strategies for the consolidation of biologically mediated events in the conversion of pre-treated lignocellulose into ethanol. RSC Adv 4, 3392–3412

Dufey A, Grieg-Gran M (2010) Biofuels production, trade and sustainable development

Brazil (2012a) Ministry of agriculture, livestock and supply. Statistical yearbook of agrienergy. https://www.agricultura.gov.br/arq_editor/file/Desenvolvimento_Sustentavel/Agroenergia/anuario_agroenergia_web_2012.pdf

Brazil (2012b) Ministry of mines and energy—Empresa de Pesquisa Energética, Plano Decenal de Expansão de Energia. https://www.epe.gov.br/imprensa/PressReleases/20120926_1.pdf

Canilha L, Chandel AK, Milessi TSDS, Antunes FAF, Freitas WLDC, Felipe MDGA, Silva SSD (2012) Department of biotechnology, school of engineering of Lorena, University of São Paulo, 12-602-810 Lorena, SP, Brazil, Received 10 October 2012; Accepted 19 October 2012

Carriquiry MA, Du X, Timilsina GR (2010) Second-generation biofuels: eco-

Cerri CC et al (2009) Greenhouse gas emissions: the importance of agriculture and livestock. Sci Agric 66:831–843

Chu S, Majumdar A (2012) Opportunities and challenges for a sustainable energy future. Nature 408:294–303

Goh CS, Tan KT, Lee KT, Bhatia S (2010) Bio-ethanol from lignocellulose: status, perspectives and challenges in Malaysia. Bioresour Technol 101, 4834–4841

Coyle W (2007) The future of biofuels: a global perspective. Amber Waves 5:24–29

DEDE (2009a) Monthly ethanol production: January 2006–August 2009. https://www.dede.go.th/dede/fileadmin/usr/bers/gasohol_documents/gasohol_2009/

DEDE (2009b) Projections of demand Supply for ethanol production. https://www.dede.go.th/dede/fileadmin/usr/bers/gasohol_documents/gasohol_2009/demand_supply_new4012010.pdfS. Accessed 19 Mar 2010

Demirbas A, Karslioglu S (2007) Biodiesel production facilities from vegetable oils and animal fats. Energy Sources, Part A 29(2):133–141

Dias De Oliveira ME, Vaughan BE, Rykiel EJ (2005) Ethanol as fuel: energy, carbon dioxide balances, and ecological footprint. Bioscience 55:593–602

Dufey A, Grieg-Gran M (eds) (2010) Biofuels Production, trade and Sustainable development. International Institute for Environment and Development, London

EPA—US Environmental Protection Agency (2010) Renewable fuel standard 2—RFS2, 2010. https://www.epa.gov/otaq/fuels/renewablefuels/index.htm

European Commission (EC) (2003) Directive 2003/30/EC of the European Parliament and of the Council. Off J Eur Union, Brussels

European Commission (EC) (2006) Green Paper, A European strategy for sustainable, competitive and secure energy. Brussels, March 8; 2006

European Commission (EC) (2008) Proposal for a directive of the European Parliament and of the Council on the promotion of the use of energy from renewable sources. Brussels, January 23; 2008

European Commission (EC) (2007) Strategic energy review—an energy policy for Europe. Brussels, January 10; 2007

European Commission, Biomass Action Plan, COM (2005) 628 final. Commission of the European Communities, Brussels

Commission E (2001) On alternative fuels for road transportation and on a set of measures to promote the use of biofuels, COM (2001) 547 provisional version. Commission of the European Communities, Brussels

Commission E (2000) Towards a European strategy for the security of energy supply, COM (2000) 769. Commission of the European Communities, Brussels

European Parliament/European Commission (2003) Directive 2003/30/EC of the European Parliament and of the Council of 8 May 2003 on the promotion of the use of biofuels or other renewable fuels for transport. European Parliament and European Council, Brussels

Ewing M, Msangi S (2009) Biofuels production in developing countries: assessing tradeoffs in welfare and food security. Environ Sci Policy 12:520–528

FAO—Food and Agriculture Organization of the United Nations (2013) FAOSTAT. https://faostat.fao.org

FAO (2009) Food security and agricultural mitigation in developing countries: options for capturing synergies. Rome, Italy. www.fao.org/docrep/012/i1318e/i1318e00.pdf

Gnansounou E, Bedniaguine D, Dauriat A (2005) Promoting bioethanol production through clean development mechanism: findings and lessons learnt from ASIATIC project. In: Proceedings of 7th IAEE European energy conference, Bergen, Norway; August 2005

Goldemberg J (2007) Ethanol for a sustainable energy future. Science 315:808–810

Goldemberg J, Coelho ST, Lucon O (2004) How adequate policies can push renewables. Energy Policy 32:1141–1146

Grahn M (2004) Why is ethanol given emphasis over methanol in Sweden? Course paper in the PhD-course Energy, Environment & Sustainability. Chalmers University of Technology, Göteborg

Greenergy International Limited (2007) Bioethanol—a greenergy perspective. London. https://www.greenergy.com

Hallock JL, Tharakan PJ, Hall CAS, Jefferson M, Wu W (2004) Forecasting the limits to the availability and diversity of global conventional oil supply. Energy 29:1673–1696

Hallock JL, Wu W, Hall CAS, Jefferson M (2014) Forecasting the limits to the availability and diversity of global conventional oil supply: validation. Energy 64:130–153

Harijan K (2008) Modelling and analysis of the potential demand for renewable sources of energy in Pakistan. PhD Thesis, Mehran University of Engineering and Technology Jamshoro

Harijan K, Memon M, Uqaili MA, Mirza UK (2009) Potential contribution of ethanol fuel to the transport sector of Pakistan. Renew Sustain Energy Rev 13:291–295

Hitam S (1999) Sustainable energy policy and strategies: a prerequisite for the concerted development and promotion of the RE in Malaysia. www.epu.jpm.my

HLPE (2013) Biofuel and food security—a report by the high level panel of experts on food security and nutrition of the committee on world food security 2013

Hoekman SK (2009) Biofuels in the US—challenges and opportunities. Renew Energy 34:14–22

https://agb.east.asu.edu/workingpapers/0207.pdf

https://www.dawn.com/2006/07/28/nat1.htmS

https://www.europa.eu.int

https://www.mpnr.gov.pk/mpnr/userfiles1/file/MPNRECCReportFinal-Updated%2009-Apr-2015.pdf

https://www.statista.com/statistics/383735/unemployment-rate-in-pakistan

Jacquet F, Bamiere L, Bureau JC, Guinde L (2007) Recent developments and prospects for the production of biofuels in the EU: can they really be "part of solution"? In: Proceedings of the conference on "biofuels, food and feed tradeoffs", Missouri, April 12–13, 2007

Jank MJ, Kutas G, Amaral LF, Nassar AM. EU and US policies on biofuels: potential impacts on developing countries. GMF study. The German Marshall Fund, Washington, DC, May 2007

Jansen JC (2003) Policy support for renewable energy in the European Union. Energy Research Centre of the Netherlands, ECN Publication, ECN-C-03-113; October 2003

Goldemberga et al., 2021Goldemberg J, Mello FF, Cerri CE, Davies CA, Cerri CC (2014) Meeting the global demand for biofuels in 2021 through sustainable land use change policy. Energy Policy 69, 14–18

Jull C, Redondo PC, Mosoti V, Vapnek J (2007) Recent trends in the law and policy of bioenergy production, promotion and use. In: Legislative study 95. Food and Agriculture Organization of the United Nations, FAO, Rome

Kaltner FJ, Azevelo GP, Campos IA, Mundim AOF (2005) Biofuelsfor transportation in Brazil. www.gtz.de/de/dokumente/en-biofuels-for-transportation-in-brazil-2005.pdfS

Karim A, Shahid Z (2018) Performance and cost analysis of conventional petrol car converted into solar-electric hybrid car. J Energy Res Technol 140(3):032009

Kerr RA (2011) Peak oil production may already be here. Science 331:1510–1511

Khan ZA (2007) Viewing biofuel (ethanol) prospects in Pakistan through a sustainable development prism. SDPI Res News Bull 14(3) July–September 2007

Kline KL, Oladosu GA, Wolfe AK, Perlack RD, Dale VH, McMahon M (2008) Biofuel feedstock assessment for selected countries. Oak Ridge National Laboratory Report, ORNL/TM-2007/224, Oak Ridge, Tennessee; February 2008

Kojima M, Mitchell D, Ward W (2007) Considering trade policies for liquid biofuels. In: Energy sector management assistance program, special report 004/07. World Bank, Washington, DC

Körbitz W, Friedrich St, Waginger E, Wörgetter M (2003) World-wide review on biodiesel production. Prepared for IEA Bioenergy Task 39, Subtask 'Biodiesel'

KTKM (Ministry of Energy, Telecommunications and Multimedia) (2004) Introduction to small renewable energy power programme (SREP). www.ktkm.gov.my/print_details.asp?Content_ID=277

Landers JN (2007) Tropical crop–livestock systems in conservation agriculture: the brazilian experience. Food and Agriculture Organization, Rome, Italy. https://www.fao.org/ag/AGP/AGPC/doc/Newpub/landers/foreword_cont.pdf

Leo-Moggie A (1996) Keynote Address. Bakun Hydro Electric Project Seminar, Kuala Lumpur, Malaysia, Mariyappan, K., 2000. Country report from Malaysia: status of renewable energy and energy efficiency in Malaysia. www.isep.or.jp/spena/2000/countryreports/malaysia.htm

Liew WH, Hassim MH, Ng DKS (2014) Review of evolution, technology and sustainability assessments of biofuel production. J Clean Prod 71, 11e29

Macedo IC, Seabra JEA, Silva JEAR (2008) Greenhouse gases emissions in the production and use of ethanol from sugarcane in Brazil: the 2005–2006 averages and a prediction for 2020. Biomass Bioenerg 32(7):582–595

Malça J, Freire F (2006) Renewability and life-cycle energy efficiency of bioethanol and bio-ethyl tertiary butyl ether (bioETBE): assessing the implications of allocation. Energy 31:3362–3380

Londo M, Deurwaarder E (2007) ECN Energy Research Centre of the Netherlands, Amsterdam, The Netherlands Perspective Developments in EU biofuels policy related to sustainability issues: overview and outlook, Wiley Inter Science (www.interscience.wiley.com), https://doi.org/10.1002/bbb.40. Biofuels, Bioprod. Bioref. 1:292–302 (2007)

Martines-Filho J, Burnquist HL, Vian CEF (2006) Bioenergy and the rise of sugarcane-based ethanol in Brazil. Choices 21:91–96

Mohamed AR, Lee KT (2006) Energy for sustainable development in Malaysia: energy policy and alternative energy. Energy Policy 34, 2388–2397

Moriartv, P., and Honnerv, D., 2016, Sustainable Energy Resources: Prospects and Policy. Chap 1 in M.G.Rasul et al (Eds) Clean Energy for Sustainable Development, Academic Press/Elsevier, London

Balat M, Balat H (2009) Recent trends in global production and utilization of bio-ethanol fuel. Appl Energy 86, 2273–2282

Nagarajan NM (2008) International symposium on agricultural and biofuel policy—presentation by Malaysia (Summary), 25 February 2008

Naik SN, Goud VV, Rout PK, Dalai AK (2010) Production of first and second-generation biofuels: a comprehensive review. Renew Sustain Energy Rev 14, 578–597

NAP (2009) Liquid transportation fuels from coal and biomass: technological status, costs, and environmental impacts. The National Academies Press (NAP), Washington, DC Economics and Policies. World Bank
OAE (2008) Agricultural Statistics of Thailand. Center for Agricultural Information, Office of Agricultural Economics, Ministry of Agriculture and Co-operatives
OCSB (Office of the cane and sugar board) (2009) Production report on sugar and sugar cane. https://www.ocsb.go.th/show_list.asp?id=13OCSBS. Accessed 15 Feb 2010
Pfuderer S, del Castillo M (2008) The impact of biofuels on commodity prices. The Department for Environment, Food and Rural Affairs, DEFRA, London
PTM (Malaysia Energy Centre) (2004) Biomass power generation & cogeneration project (BioGen). https://www.ptm.org.my/New_BioGen_Web_II
PTM (Malaysia Energy Centre) (2008) Energy smart special focus: mitigating the adverse impact of climate change through sustainable use of energy. https://www.ptm.org.my
Ramchandra TV, Joshi NV, Subramanian DK (2000) Present and prospective role of bioenergy in regional energy system. Renew Sust Energy Rev 4, 375–430
REN21—Renewable energy policy network for the 21st century (2013) Global Status Report, 2013. https://www.ren21.net/REN21Activities/GlobalStatusReport.aspx
Renewable Energy Network for the 21st Century (REN21) (2008) Renewable 2007 Global Status Report. Paris: REN21 Secretariat and Washington, DC: Worldwatch Institute
Renewable Energy Network for the 21st Century (REN21) (2007) Renewable 2006 Global Status Report. Paris: REN21 Secretariat and Washington, DC: Worldwatch Institute
RES-D—European Union (2009) Directive 2009/28/EC of the European Parliament and of the Council of 23 April 2009 on the promotion of the use of energy from renewable sources amending and subsequently repealing Directives 2001/77/ EC and 2003/30/EC. Off J Eur Union L140/16 of 5.6.2009
Robertson F (2003) Sugarcane trash management: consequences for soil carbon and nitrogen. BSES and CRC Sugar, Townsville
Schnepf R (2006) European Union biofuels policy and agriculture: an overview. Congr Res Serv Rep, RS22404, Washington, DC, March 16; 2006
Seelke CR, Yacobucci BD (2007) Ethanol and other biofuels: potential for US–Brazil energy cooperation. CRS Report for Congress, Order Code RL34191, Washington, DC, September 27; 2007
Silalertruksa T, Gheewala SH, Sagisaka M (2009) Impacts of Thai-bioethanol policy target on land use and greenhouse gas emissions. Appl Energy 86:S170–S177
Sorda G, Banse M, Kemfert C (2010) An overview of biofuel policies across the world. Energy Policy 38(6977e):6988
Sriroth K, Piyachomkwan K (2008) Cassava ethanol technology and growth in Thailand
Tan KT, Lee KT, Mohamed AR (2008) Role of energy policy in renewable energy accomplishment: the case of second-generation bioethanol. Energy Policy 36:3360–3365
Thamsiriroj T, Murphy JD (2009) Is it better to import palm oil from Thailand to produce biodiesel in Ireland than to produce biodiesel from indigenous Irish rape seed? Appl Energy 86:595–604
Timilsina GR, Mevel S, Shrestha A (2011) Oil price, biofuels and food supply
TTDI (Thai Tapioca Development Institute) (2004) https://www.tapiocafeed.com/tapioca/breed.htmlS. Accessed 12 Mar 2010
UNEP (2008) Towards sustainable production and use of resources: assessing biofuels. UNEP DTIE, Paris
UNFCC (United Nations Framework Convention on Climate Change) (2008) Clean development mechanism 2008. https://unfccc.int/resource/docs/publications/08_cdm_in_brief.pdf
van Thuijl E, Deurwaarder EP (2006) European biofuels policy in retrospect. ECN, Petten
Wang G, Lixin Z, Yanli Z, Yujie F, Huba EM, Dongsheng L, Shizhong L, Dehua L, Mang HP. Liquid biofuels for transportation: Chinese potential and implications for sustainable agriculture and energy in the 21st century. Assessment Study, Institute of Nuclear and New Energy Technology, Tsinghua University, Beijing; February 2006

WBGU (German Advisory council on Global Change) (2010) Future bioenergy and sustainable land use. Earthscan, London

Welfle A, Gilbert P, Thornley P (2014) Securing a bioenergy future without imports. Energy Policy 68, 1e14

Wiesenthal T, Leduc G, Christidis P, Schade B, Pelkmans L, Govaerts L et al (2009) Biofuel support policies in Europe: lessons learnt for the long way ahead. Renew Sustain Energy Rev 13:789–800

World watch Institute (2006a) Biofuels for Transport: global potential and implication for sustainable energy and agriculture, London

World watch Institute (2006b) State of the world 2006: special focus: China and India. A world watch institute report on progress toward a sustainable society. World watch Institute, Washington, DC, January 7; 2006

Yacobucci BD, Schnepf R (2007) Ethanol and biofuels: agriculture, infrastructure, and market constraints related to expanded production. CRS Report, Congressional, Research Service Washington, DC, RL33928, March 16; 2007

Yan J, Lin T (2009) Biofuels in Asia. Appl Energy 86:S1–S10

Yang B, Lu Y (2007) Perspective the promise of cellulosic ethanol production in China. J Chem Technol Biotechnol 82:6–10

Chapter 10
Environmental Impacts of Ethanol Production System

Nadia Sharif, Neelma Munir, Maria Hasnain, Shagufta Naz, and Muhammad Arshad

Abstract The diminishing fossil fuel assets and their expanding costs have prompted a worldwide scan for alternative energy assets nowadays. Sustainable energy sources empower change of ecological protection and are an imperative component of sustainable improvement. Utilizing biofuel positively affects nature, the reasons being it provides the energy safety, improved transport facilities and decreases the emission of exhaust gases. The encouraging feature of biofuel is the ethanol production from sustainable energy sources. Dissimilar to gasoline, ethanol is biodegradable and less toxic to environment and it rapidly breaks down into innocuous affluences if spilled. In the fuel market, diverse ethanol-gasoline mixtures are available. In spite of the fact that ethanol from grain has been reprimanded for its net energy balance, noteworthy arable land, water necessities, and natural effects, for example, N_2O pollution, biodiversity loss, soil erosion, and high volatile organic compounds. However, bioethanol obtained from biomass is perceived as a suitable substitute for transportation fuel. Sugarcane and corn like feedstocks of ethanol production engross CO_2 as they develop and could offset the CO_2 produced when ethanol is generated and burned so it is recognized as atmospheric carbon–neutral agent. Hence, bioethanol absolutely has potential to reduce fossil resource use and greenhouse gas emanations, if resources such as sugarcane and switchgrass are used.

Keywords Biofuel · Bioethanol · Environmental impacts · Sustainable energy · Fermentation

---

N. Sharif
Department of Biotechnology, Women University Mardan, Mardan, Pakistan

N. Munir (✉) · M. Hasnain · S. Naz
Department of Biotechnology, Lahore College for Women University, Lahore, Pakistan
e-mail: neelma.munir@yahoo.com

M. Arshad
Jhang-Campus, University of Veterinary and Animal Sciences, Lahore, Pakistan

M. Arshad (ed.), *Sustainable Ethanol and Climate Change*,
https://doi.org/10.1007/978-3-030-59280-6_10

Energy is the dominant contributor to climate change, accounting for around 60 percent of total **global** greenhouse gas emissions. Sustainable energy sources empower change of ecological protection and are an imperative component of sustainable improvement. Utilizing biofuel positively affects nature, the reasons being it provides the energy safety, improved transport facilities and decreases the emission of exhaust gases. Ethanol is biodegradable and less toxic to environment and it rapidly breaks down into innocuous affluences if spilled. Bioethanol absolutely has potential to reduce fossil resource use and greenhouse gas emanations, if resources such as sugarcane and switch grass are used.

10.1 Introduction

In view of ecological pollution and fossil fuel exhaustion, biofuel production has pulled in impressive considerations. Hydrogen and ethanol are considered to be the auspicious energy bearers because of their sustainability and ecological amicability. Ethanol production by photosynthetic plants using CO_2 and sun solar energy gives an alluring answer for sustainable production of green fuels. The scaling up procedures of ethanol production were generally undermined or even crushed by biocontaminations, which confined biomass or items aggregations of feedstocks. Besides this, it is of significance to create dependable biocontamination controlling methodologies, for advancing ethanol production in huge scales (Lin et al. 2012).

Dark fermentation is an alluring path for ethanol and hydrogen production since it can use different types of raw materials of organic nature as substrate. A framework for ethanol production and related products from lignocellulosic biomasses, especially from sugarcane bagasse and straw is provided, however, not constrained thereto, incorporated with traditional procedures for the production of ethanol. First era ethanol production from two crops, sugarcane and sorghum in four modalities, possibly consider the supportability level of every methodology (Lin et al. 2012; Liu et al. 2013).

The use of lignocellulosic biomass for ethanol production is an auspicious energy hotspot for the restricted raw petroleum. There are basically two procedures elaborated in the transformation: cellulose hydrolysis in the lignocellulosic biomass for reducing sugars production, and sugars fermentation to ethanol. In the light of current innovations, ethanol production from lignocellulosic feedstocks is abstemiously expensive, whereas hydrolysis procedure cost and low yields are the fundamental difficulties. Extensive research exertions have been applied to enhance the lignocellulosic feedstocks hydrolysis. Pretreatment of lignocellulosic feedstocks to evacuate lignin and hemicellulose can fundamentally improve the cellulose hydrolysis. Streamlining of the cellulose compounds and the enzyme loading can likewise enhance the hydrolysis. Concurrent fermentation and saccharification viably evacuate glucose, which is a cellulase activity inhibitor, therefore, expanding the yield and rate of cellulose hydrolysis (Sun and Cheng 2002).

Very High Gravity (VHG) innovation was utilized on industrial scale to create ethanol from molasses and also by products formation estimation. To reduce the ethanol production cost from molasses at commercial scale, VHG with continuous air flow is a sustainable technique. Another choice for oil-based fluid transportation is the ethanol production by utilizing cellulosic biomass. Due to low density of cellulosic feedstocks, expansive scale manufacturing of cellulosic ethanol should be avoided. Trials have demonstrated that pelleting by ultrasonic vibration-assisted method has resulted in packing low-density crude biomass into high-density pellets. This procedure is affected by temperature. Solid acid catalysts have the ability to hydrolyze cellulose and can bring down reaction times by simple procedures and can be recovered and reused easily. A glycerol-based carbon acid catalyst created at CSIR-IICT performed well in acid catalyst reactions and subsequently this investigation was embraced to assess the acid catalyst for biomass hydrolysis (Goswami et al. 2015; Zhang et al. 2016; Arshad et al. 2017).

Factors causing the wide variety in the appraisals of the net energy estimation of maize ethanol in past studies were identified and a steadier estimate was established. It is inferred that the net energy estimation of maize ethanol has turned out to be certain because of innovative advances in ethanol conversion and increased proficiency in cultivate generation (Shapouri et al. 2003; Solomon et al. 2014).

Energy outputs of ethanol production by utilizing switchgrass, wood biomass, and corn were each not as much as the particular fossil energy inputs. Sugarcane cultivation upgrading and picking procedures, e.g., green cane manufacturing alongside coordinated use of biomass would benefit in decreasing the ecological effects of the fundamental harvests got from sugarcane like ethanol and sugar. The probable ways could be reduced by around 38% on climate change, acidification is reduced by 60%, photo-oxidant formation is reduced by 90%, particulate matter formation is reduced by 63%, and fossil depletion is reduced by 21% (Silalertruksa et al. 2017). Studies demonstrate that the decision of apportionment strategies has the best effect on fuel ethanol net energy. The system development method can totally eradicate provision strategies in the forefront ethanol making system from corn grain (Kim and Dale 2002). For the development of economical and renewable biofuel from energy grasses or lignocellulosic polysaccharides, a few pathway engineering methodologies have been accounted for. The generation of novel metabolic pathways by coalescing genes from particular entities in selected host to manufacture products is among them. In the post-genomic era, synthetic biology pursues to construct and model biological machineries, roles, and entities that don't exist in environment or to promote current biological systems to execute new roles. Along these lines, metabolic engineering methodologies in synthetic biology are congregating with synthetic genomics. The probable ways could be using synthetic biology, chemical engineering, imaging, chromatography, bioinformatics, protein engineering, enzyme technology, fermentation technology, and genome sequencing (Kumar et al. 2017).

If the accessible feedstock assets are used sustainably and implementable strategies are drawn in apt direction, Pakistan has a significant potential for biodiesel production. To find non-conventional alternate resources of energy and to meet the energy demands various challenges need to be stated. Such challenges are infrastructure advancement, regionalized sort of power delivery system, commercialization, marketplace expansion, sentience, monitoring, appropriations, administrational contribution, technology transfer, and assessment. In order to integrate and control at national level, a comprehensive policy is also needed. With the advantage of great diversity of feedstocks Pakistan is most suitable for biodiesel fabrication. Pakistan has an enormous potential for biodiesel production from jatropha, plants seed oil, and microalgae have enormous potential for biodiesel production sources of Pakistan but practical applications and consideration is required. Due to lack of financial resources, absence of policies to promote and develop alternative energy resources Pakistan is facing challenges in the commercialization of biodiesel (Shah et al. 2018). Liquid biofuel industry in Pakistan may result in saving our foreign exchange and environment (Mirza et al. 2009).

Sustainable improvement expects to accomplish financial thriving while at the same time keeping up and maintaining environmental protection. To achieve sustainability in practice, society must help the existing methodologies and activities. The execution of policies to stimulate the fabrication and utilization of biofuels to reduce Greenhouse Gas (GHG) emissions and environmental impacts among the world's most noteworthy energy consumers is an example of such sustainability strategies (Cherubini et al. 2009). It is required as the longer we wait to reduce our emissions, the more expensive it will become (Griggs and Noguer 2002; Allen et al. 2014).

10.2 Impact of Ethanol Production Technologies

Ethanol is produced by fermentation, a chemical reaction induced by catalysts enzymes produced by living cells. There are a number of technologies that use different raw materials containing simple sugars that can be directly subjected to fermentation. By using current technologies for the production of bioethanol, the least expensive has been found in the sugarcane in Brazil and starch crops in Europe (Ryan et al. 2006). By and by the production of ethanol to a great extent focus on utilizing surplus wastes and residual biomass of crops like maize and wheat stalks, energy crops like switchgrass, rapidly developing plants, e.g., willow and poplar, waste paper and packing measureable, oats in type of grain inadmissible for utilization, etc. (Kim and Dale 2004; Azizov 2006).

However, despite different methods checked for their successful application in ethanol production at lab scale, they are not in practice. The aggregate probable bioethanol production from waste is around 16 times high as compared to current world ethanol production (Silveira et al. 2015). Production expenses of bioethanol vary and are subject to the costs of raw constituents, the technique for production, the degree of refining attempted, and the auxiliary use of wastes and bioproducts. Contingent on the raw materials processing degree, the process (i.e., the ratio of energy contents and energy supplied for production) energy output is different. By using different materials, the energy output of ethanol production varies from 1.7 to 3.8 and is found to be inversely proportional to the processing time of the materials subjected to fermentation. Subsequently, the current huge enthusiasm for biofuel production advances utilizing waste materials. Besides, the ceaseless generation process made out of enzymatic hydrolysis, thermos pressure hydrolysis, fermentation and ethanol dewatering is proposed, which is described by high level of heat recuperation and low production cost. Another option for what's to come are biorefinery multisystem producing food, solvents, fuels, and plastics from waste biomass and including lactic acid and ethanol fermentation (Ohara 2003; Marszałek and Kamiński 2008).

10.2.1 Conventional Corn Starch Ethanol

It is hard to interface the ecological effects of corn production to particular unit processes. All in all, traditional ethanol production gives a 13–19% decline in all out total GHG emissions over gasoline production. GHG emissions from corn starch ethanol generation, about 34–44% comes from the agrarian procedures of cultivating corn. Ethanol amenities could intense their GHG diminishments to 28–39% if they practice natural gas as an energy cradle, which is the situation with numerous present existing ethanol amenities (Wang et al. 2007; Yeh et al. 2008).

About 33% of the natural gas thermal energy is utilized to dry the distiller's grains, making it the huge single supporter of GHG emissions. Be that as it may, with expanding flammable gas costs, a few plants are investigating supplanting natural gas and utilizing biomass as their fuel source. Add up to GHG diminishments for utilizing elective fuel cradles are 39–52% over production of gasoline. Nevertheless, while traditional ethanol conveniences utilize coal as their energy cradle, entire ecological advantages vanish and miserably, with the increasing expenses of natural gas a few plants are thinking about doing this change to coal (Wang et al. 2007). Additional regular natural apprehension encompasses the ethanol diligence is their water necessities. Normal water use for dry mill ethanol production is around 4–4.7 gallons H_2O/gallon ethanol. These numbers have munificently diminished after some time and were initially as high as 15 gallons H_2O/gallon ethanol (Shapouri et al. 2003).

Water reuse is of key significance in minimizing total water utilization. Nevertheless, reusing water cannot definitely diminish the current normal water use because of the subsequent exceptionally concentrated salts. Recovery of evaporated water is required to maintain the cycle. In spite of declining per gallon ethanol prerequisites, water required by the ethanol industry keeps on ascending because of increasing numbers of plants and increasing plant capacities (Kelly et al. 2007). Management tillage was practiced done for more than 40 years and showed real environmental progresses over erosive open plowing of the past. Nevertheless as to the magnitude its consumption on cornfields, the U.S. Department of Agriculture stated in 2015 that this tillage applies were applied on more than 30% of all U.S. corn acres in 2010–11, generally exterior the Corn Belt. This percentage is lower than wheat, soybeans, or cotton. An agricultural economist—Brian Wright, at the University of California at Berkeley—has noted that real corn prices have nearly doubled since the ethanol mandates of 2005–2007.

10.2.2 Cellulosic Ethanol

Cellulosic ethanol has a huge environmental potential, this potential is because of three primary components sketched out in. To begin with, upon combustion, biomass discharges a same amount of CO_2 as it is already expended and in this way no new atmospheric carbon is introduced. Second, cellulosic energy crops have brought down fossil fuel inputs than conventional row crops and finally the co-products of cellulosic ethanol can be utilized to meet plant energy prerequisites, wiping out the requirement for fossil fuel energy sources. The size of the potential just increments with upgrades in feedstocks. The total reduction in GHG due to petroleum fuel varies. About 85% reduction in GHG emanations per gallon of ethanol used to dislodge an energy-equivalent measure of fuel is predicted. A life-cycle examination of corn feedstock resulted that for every kilometer distance utilizing 100% ethanol, GHG emissions would diminish by 254 g/km, which equates to a 106% reduction of gasoline use. When apprehensive with particular GHG, cellulosic ethanol may expand emissions, contingent upon the feedstock.

The measure of CO_2 delivered in the state is profoundly subject to the conveyance method of Stover. Despite that, corn grains likewise require feedstock transportation and in this way cellulosic ethanol still ought to decrease add up to CO_2 emanations over gas and regular starch ethanol. Notwithstanding the way that the ozone-depleting substance diminishments are noteworthy over traditional ethanol creation, biochemical cellulosic production improves with regard to water utilization. Cellulosic ethanol delivered by means of dilute acid hydrolysis is anticipated to at first requires around 9.5 gallons H_2O/gallon ethanol (Suh and Suh 2010; Creutzig et al. 2015).

In any case, as occurred for regular ethanol production, water utilization for cellulosic ethanol production will probably decline as the industry develops. The anticipated water use for the "nth" cellulosic ethanol plant will probably be between two to six gallons H_2O/gallon ethanol. In the event that cellulosic ethanol production can accomplish these diminishments in water utilization, it will be keeping pace conventional ethanol production (Pate et al. 2011; Li et al. 2016).

The environmental advantages of gasification over biochemical and conventional ethanol production are uncertain at present, generally because of deficient data. There is restricted research on the emissions from burning distinctive feedstock in a gasifier, and there are concerns about increased nitrous oxide emissions from gasifying a few feedstocks, for example, dried distiller's grains. Nonetheless, environmental advantages to utilize biomass over natural gas for gasification, and environmental advantages to utilizing cellulosic feedstocks over corn starch. In this manner, gasification will probably yield environmental boosts over conventional ethanol facilities; however, their advantages contrasted with biochemical cellulosic production are obscure. The water utilization for gasification is also dubious; it could be as low as 2 gallons H_2O/gallon ethanol. Gasification will probably fall in a comparative range to future anticipated biochemical water utilization of 2–6 gallons water/gallon ethanol. As some time recently, while these water uses may appear to be reasonable on a plant basis, the combined water use of the industry will remain a concern (Koh and Ghazoul 2008).

10.2.3 Comparison of Environmental Impacts

First or best alternate to guarantee widespread and impartial contribution of feedstocks in biofuel production could not only be provided by regulatory tactics to standards and certification. For the environmental amenities outlays may also signify an appliance for reassuring acquiescence with sustainable production policies. Similarly foods, agricultural crops, and biofuel feedstocks should be treated. In general, the environmental apprehensions of bioethanol feedstocks are same as for the effects of amplified agronomic products; subsequently, sustainability measures should be applied constantly to all crops (Romano 1998).

As they can for extensive agronomic products, good agrarian practices like agriculture conservation decrease the carbon imprint and minimize the environmental effects of biofuel. Grasses or trees like perennial feedstock crops help in improving the degraded or marginal land as well as diversify the production systems. Improved international significances of biofuel development are required in result of local governmental strategic plans. Realistic and achievable biofuel mandates and targets are only possible through international dialogues on existing mechanisms. Therefore, there is a need of good practices with accessible knowledge. Issues related to natural resource management like land, water soil, and biodiversity are more prone to life cycle analysis. Overall sustainable bioenergy development among the zones wherever good practices are expected to discourse including agrochemical management, nutrients, crop, soil and water protection, landscape conservation and biodiversity, harvesting, processing and distribution and also energy and water management.

Conservation agriculture is the practice that is helping the people to accomplish profitable and sustainable agriculture for rural and farmers publics by using minimal soil disturbance, diversified crop rotations and perpetual organic soil cover (Doornbosch and Steenblik 2008).

In accordance with the current quintessence on carbon storing and on inventions that reduce energy power, it happens to be predominantly appropriate. In eminences where labor is sporadic and conservation of soil moisture and fertility is needed at that places tactic validates are amenable. Some strategies that are applied to resolve the issues include, for example, mechanical soil tillage interventions, application of nutrients of organic or mineral origin, and agrochemicals at an optimum level, as well as measures, are taken that do not intrude biological processes. Athwart agro-ecological regions and cultivation systems, the conservation agriculture has been appeared in ersatz operational (Dixon et al. 2012). Great farming and forestry practices may perhaps exceedingly decrease ecological incidentals allied using sustainable intensification and plausible expansion at forest edges. When bioenergy feedstocks form part of the mix, the practices in light of agro-silvo-pasture-livestock amalgamation are considered additionally. Though miscellaneous ecological effects of bioenergy encroachment do not diverge applicably from those of different types of agronomic, the need is in what way they could be better determined in field accomplishments. Prevailing ecological effect evaluation methods and key natural appraisals offer a attired beginning phase for breaking down the biophysical factors. There is like manner exists an abundance of specialized learning drawn from agrarian improvement in the midst of the previous sixty years. Different responsibilities of bioenergy setting incorporate expository outlines for bioenergy and sustenance safety also deal with the total natural effects, including air contamination, biodiversity misfortune, pesticide lethality, over the top manure utilize, and soil acidification; what's more, chip away at social and ecological manageability criteria, including unfriendly effects on biodiversity, supplement draining, restricts on deforestation, soil disintegration, and competition with food production (Cherubini and Jungmeier 2010).

The biofuel section is depicted by a broad assortment of accomplices with grouped interests. This, joined with the quick headway of the segment, has provoked an expansion of activities to ensure manageable bioenergy advancement. Standards, criteria, and prerequisites are under idea, alongside consistence instruments to measure execution and direct improvement of the division. Bioenergy Partnership's teams on ozone-depleting substance strategies and on supportability, and the round table on economical biofuels, consider as a genuine piece of these, together with various other open, private and non-profit tries. This attired variability proposes a strategy for fitting the diverse techniques might be required, particularly arranges approach and centers on fortifying more biofuel production (Sitnikov 2013).

Utilizing ethanol over gasoline prompts decreases in GHG emissions. The diminishments are modest for conventional ethanol production; however, by making little modifications in the production procedure, e.g., utilizing natural gas, plants can accomplish the higher end of their GHG reduction potential. Cellulosic ethanol production likely surpass conventional corn starch reductions, however, evaluates for thermochemical production are meager. Furthermore, gauges for biochemical cellulosic ethanol are extremely encouraging, with 85% or more noteworthy reductions in greenhouse gases. Caution for cellulosic ethanol production is the potential for individual greenhouse gases to increase like nitrous oxide, even with general reductions in emission. For each of the three technologies, water use will probably go in the vicinity of two and six gallons H_2O/gallon ethanol. Cellulosic ethanol water use diminishes at an indistinguishable rate from water utilization has over the most recent ten years of momentum ethanol industry. It seems to control ten years for cellulosic ethanol reduction to its foreseen scope of two to six gallons H_2O/gallon ethanol (Koh and Ghazoul 2008).

Though a number of ideashave been proposed by industrial countries but to date none have yet been tried, especially in conjunction with government bolster designs (Doornbosch and Steenblik 2008).

The stint canons derive thorough frameworks in lieu of estimating factors contrary to determined standards, whose inability to go along would keep a state as of exchanging its item. These globally concurred frameworks as of now exist for a scope of food security, concoction, and human well-being themes. From one perspective, given that most natural effects of biofuels are unclear from those of expanded farming creation by and large, it could be guaranteed that equivalent measures ought to be connected in all cases. Besides, confining area utilizes change could dispossess open doors for creating nations to benefit from expanded interest for rural wares. Then again, there are additionally solid sentiments that agrarian makers and strategy creators ought to gain from prior errors and stay away from the negative ecological effects that have gone with farming area transformation and intensification previously (Ammann 2013).

Answers for this issue will require cautious exchange and transaction among nations if the joined objectives of agricultural efficiency development and natural manageability are to be expert. A beginning stage may be found by setting up best practices for a reasonable biofuels production, which would then have the capacity to similarly help change cultivating, rehearses for non-biofuel crops. In time, and joined by limit building endeavors for the nations that need it, more thorough norms and certification frameworks could be set up. One decision to research could be installments for natural administrations in blend with biofuel generation. This instrument would repay agriculturists for giving specific natural administrations utilizing generation strategies that are ecologically more practical. Installments could be connected to consistence with gauges and certification plans concurred at the universal level. Installment plans for ecological administrations, albeit testing and confused to execute, could constitute a further instrument to guarantee that biofuels are delivered in a supportable way (FAO 2008).

10.3 Impact of Ethanol Dehydration Methods

Ethanol got amid ethanol fermentation and rectification has around 95% ethanol. Production of anhydrous ethanol utilized for fuel purposes requires defeating the hindrances of a positive homoazeotrope. Presently, the most essential ethanol dehydration techniques utilized as a part of the world industry comprise azeotropic distillation and dehydration on molecular sieves. Another choice to the conventional strategies for ethanol lack of hydration is Pervaporation (PV) or Vapor Saturation (VP) the new age of division systems (Chemtech 2004; Kaminski and Marszalek 2006). In the midst of pervaporation, a fluid stream is isolated on a semi-penetrable film into two streams: a vaporous saturate which is advanced with water and fluid retentate which is improved with ethanol (Marszałek et al. 2009). A correlation of the expenses of ethanol dehydration by numerous techniques in a bigger system indicates that operating expenses of the membrane techniques are smaller by half than that of other dehydrating methods (Sander and Soukup 1988, 1991).

PV is defensible when at the inlet water concentration in the system is under 10%. In the event that still higher product dehydration is normal, at that point significantly greater membrane surface and higher weight diminishment in favor of saturate is required. Cost of ethanol drying out abatements with an expansion of saturation transition and mass portion of ethanol in pervade. It is plausible to distinguish concrete trends in the development of PV in ethanol dehydration industry. The PV establishment can be an autonomous, last phase of dehydration after fermentation process (Mathur and Medintz 2017).

There are many examples of hybrid processes of PV with distillation. Such hybrid processes empower lower energy demand and not to utilization of added substances. Nevertheless, investment outlays like process complexity and high membrane prices. It has extensive applications in industry, depending on high process productivity it will decrease the membrane cost. The hybrid frameworks will achieve financial favorable circumstances at long haul forms; however, they are not productive because of little ethanol drying out frameworks. Next to working of new ethanol dehydration systems based on hybrid processes of distillation PV, producers offer also additional use of the PV in effective existing establishments (Chemtech 2004).

The PV module can be set between the refining and azeotropic section, this will be a two-fold increment of effectiveness and related reduction of energy price, dehydration costs, more efficient utilization of the current system, and a possibility to control the PV module. Thus, the PV module can be associated with the effectively existing dehydration on molecular sieves, this will cause an expansion of process effectiveness and the amount and quality of water removed, reduction of product recirculation degree and energy utilization. Basically, the final application involving setting PV among refining and adsorption on sub-atomic strainers can acquire the future financial advantages while ethanol being diminishing (Karthikeyan et al. 2009).

10.4 Reduction in Greenhouse Gas Emissions

Concentration of greenhouse gases in the atmosphere has increased to levels unprecedented on earth in 800,000 years (Woodward et al. 2014). Various policy-makers accepted that the superseding of non-renewable energy sources with powers produced as of feedstock might have positive and significant environmental variation impacts. These sources produce lower levels of the ozone harming substances that improve an Earth-wide temperature boost. Bioenergy yields can lessen or counterbalance ozone-depleting substance emanations by clearly expelling CO_2 from the air as they develop and putting away it in biomass and soil.

Depending upon the procedures used to process the feedstock for fuel, a couple of products can even create more ozone harming substances than do petroleum derivatives. For example, nitrous oxide, an ozone-depleting substance with an Earth-wide temperature likely boost 300 times more inimitable as of CO_2, is released from N_2 manures. Furthermore, ozone-depleting substances are released at various phases in biofuel production (FAO 2008).

The potential of different biofuels to diminishing non-renewable energy source varies extensively. It depends on factors, for instance, feedstock attributes, creation area, rural practices, and energy transformation procedures. Assorted biofuels likewise perform distinctively to the extent their commitment to diminishing ozone-depleting substance emanations (FAO 2008).

Biofuels are just the substitutes for alleviating GHG emanations. Contingent upon the strategy intents, different alternatives may demonstrate further cost-effective, as well as assorted types of sustainable power source, expanded vitality efficiency and protection, and lessened emanations from deforestation and land debasement. In spite of that the effects of expanded biofuel generation on ozone-depleting substance emanations, biodiversity, land and water change extensively athwart nations, biofuels, sources and creation hones, at hand is a solid and incite requirement aimed at orchestrated ways to deal with life-cycle examination, ozone harming substance adjusts, and supportability criteria. Ozone-depleting substance adjusts are not constructive for all feedstocks. For environmental change devotions, venture ought to be headed for crops that the most significant constructive ozone-depleting substance adjusts with the least ecological and social expenses (FAO 2008).

Life cycle investigation is the expository device used to compute ozone-depleting substance adjusts. The greenhouse gas adjust is the aftereffect of an examination between all emanations of ozone-depleting substances, all through the generation stages and utilization of a biofuel and all the ozone-depleting substances discharged in creating and utilizing the comparable energy measure of the separate petroleum derivative. This entrenched, however intricate, technique deliberately investigates every part of the esteem fastens to appraise ozone harming substance outflow. Ozone-depleting substance adjusts changes differentiate extensively among harvests and areas, contingent upon feedstock generation strategies, transformation advancements, and utilization. Data sources, for example, nitrogen compost and the kind of power age used to change over feedstocks to biofuels may bring about broadly fluctuating levels of ozone harming substance discharges and furthermore contrast with one locale then onto the next (FAO 2008).

Production of first-age biofuels from current feedstocks could reduce 20–60% petroleum products dependence. Transformation of large feedstock fields to create ethanol and biodiesel in many countries releases about 17 fold CO_2 as those biofuels spare yearly by superseding fossil fuels (Fargione et al. 2008; Rhodes 2014).

A study performed by Righelato and Spracklen (2007) the carbon emissions from the bioethanol produced from different feedstocks grown on current cropland was evaluated. It was discovered that, for each situation, over a 30-year time frame more carbon would be sequestered by changing over the cropland to timberland. It was recommended that if the goal of biofuel bolster approaches is to relieve an unnatural weather change, at that point fuel efficiency and woods preservation and reclamation would be more successful choices.

Doornbosch and Steenblik (2008) assessed that decreasing ozone harming substance emanations by means of biofuels costs substantially higher than the market cost of CO_2 equal balances. Enkvist et al. (2007) reported the clear measures to diminish energy utilization, for example, better protection of new structures or expanded efficiency of warming and aerating and cooling frameworks. The estimation of ozone-depleting substance and vitality adjusts of bioethanol is mind-boggling (Ryan et al. 2006). Bioethanol produced from Brazilian sugarcane has a better well-to-wheels energy balance and CO_2 emissions savings over 85% (Ryan et al. 2006; Demirbas 2007). Notwithstanding the effects of feedstock creation on ozone harming substance outflows, biofuel handling and appropriation can likewise have other ecological effects (Dufey 2006).

In any case, to the degree that biofuels can supplant customary biomass, for example, fuelwood and charcoal, they additionally grasp prospective for emotional enhancements in human wellbeing, especially of ladies and youngsters, over diminished respiratory maladies and death affected by indoor air contamination. Now and again, national directions expect shippers to guarantee the practical development of rural land, the assurance of normal environments, and a base level of CO_2 reserve funds for biofuels. A few countries and territorial associations have suggested that net ozone harming substance adjusts from biofuels should be in the extent of 35–40% not as much as that of petroleum. A watchful investigation of such concerns is vital for all partners, as a reason for speculation and creation choices and guaranteeing the attractiveness of their items (FAO 2008).

Two of the primary main impetuses behind approaches advancing biofuel advancement have been worries over vitality security and a want to diminish ozone-depleting substance outflows. The scope of assessed petroleum derivative adjusts for cellulosic feedstocks are even more extensive, reflecting the vulnerability in regard to this innovation and the decent variety of potential feedstocks and creation frameworks. Nevertheless, evaluating the consequence of a biofuel on ozone harming substance emanations needs investigation of discharges all through the span of the life cycle of the biofuel right off the bat planting and assembling the feedstock at that point preparing the feedstock into biofuel (Baxter et al. 2009).

Land utilize changes related with extended biofuel production can have a noteworthy effect. For instance, changing over timberland or agrarian yields uprooted by biofuel sources somewhere else can discharge large amounts of carbon that take a very long time to recuperate through the emanation decreases accomplished by substituting biofuels for petroleum derivatives (FAO 2008).

10.5 Other Impacts (Acidification, Toxicity, Impacts on Biodiversity, Soil, Water, and Land Use Aspects)

10.5.1 Soil Utilizes Perspectives

Both lands modification and intensification of agronomic crops cultivation on prevailing croplands can have significant unfavorable effects on soils, yet these effects similarly concerning any crop basically hinge on cultivating methods. By the removal of permanent soil cover, the incongruous cultivation procedures can increase soil erosion and decrease soil organic matter. The GHG emission increases, loss of soil carbon and soil nutrient contents is reduced through evacuation of plant residues. Similarly, some other management practices such as crop rotations and conservation tillage could help in improving environmental eminence by minimizing adverse environmental effects with the advantage of increased bioethanol feedstock production. Soil quality could be improved in terms of increasing organic carbon levels and soil cover by growing perennials instead of annual crops. Distinctive feedstocks differ from their nutrient demand, level of land provision they need and by their oil impacts. Through recycling the nutrients from distillery wastes and sugar-mill the soil quality is kept upward. Reuse of deposits for nutrients recycling and soil fertility maintenance require extensive production systems (Wilhelm et al. 2007; Doornbosch and Steenblik 2008). Less intensive management and less fossil energy inputs are needed for perennial lignocellulosic crops like grasses, eucalyptus, willow, or poplar. Such crops could increase the soil quality and carbon after cultivation on poor quality land (FAO 2008).

10.5.2 Water Utilizes Viewpoints

In several specific situations, rather than land, the water scarcity acts as the vital limiting factor for biofuel biomass production. Around 70% of freshwater reserved universally is used for agricultural purposes. Water resources for agriculture are ending up progressively rare in numerous countries because of amplified competition with industrial or domestic uses. Besides, the probable effects of climate change as far as of abridged rainfall and runoff in certain significant producer areas will abode more compression on already short resources (De Fraiture et al. 2008).

Crops now used for biofuel production having generally high water necessities at marketable yield echelons and are hence most appropriate to l tropical zones. In Brazil, seventy-six percent of sugarcane production is in tropical areas, and in the USA 70% of maize production is rain-fed. Even perennial plants that are cultivated in semi-arid areas on peripheral or tainted plots might need irrigation while dry and seasons. Additionally, the feedstocks processing into bioethanol can utilize huge amount of water, chiefly for seeds and plants washing and for evaporative cooling. In any case, local water resource balance is maintained through the irrigated production of key biofuel feedstocks. In northeastern Brazil and southern and eastern Africa various irrigated sugar-producing areas are now working adjacent the hydrological confines of associated river basins. On the basis of water reservoirs and land areas, probable extension of irrigated zones may appear high in certain areas; to assure water and land conveyances that might not imitate through commercialized schemes; the definite opportunity for amplified fuel production beneath irrigated situations on prevailing lands is inadequate. Correspondingly, extension may be constrained by advanced peripheral charges of water storing and land acquisition (Chum et al. 2014).

Water quantity as well as quality is affected by biofuel crops production. For instance, conversion of woodlands or pastures into maize fields could cause problems like infiltration into groundwater from excessive fertilizers solicitation, excess nutrient runoff, sedimentation, and soil erosion. Oxygen starved "dead zone" in the Gulf of Mexico is due to excessive nitrogen in the Mississippi river system where various forms of marine life cannot subsist. Ethanol and biodiesel production consequences in organically polluted water that, if comes out unprocessed, might upsurge eutrophication of superficial water bodies. Conversely, already practiced wastewater management technologies can covenant viably with wastes and organic pollutants. More than 90% biological oxygen demand of wastewater is reduced by fermentation systems, consequently, water could be recycled and reused as well as the methane could be apprehended in the treatment scheme and castoff in energy cohort. As reverences the dissemination and stowing phases of the cycle, since biodiesel and ethanol are decomposable, the latent for adverse influences on water and soil from drip and spills is abridged related with fossil fuels. Chemicals and pesticides washed into water bodies and deleteriously affect water quality (Tilman et al. 2006).

10.5.3 Biodiversity Impacts

Biofuel production impacts agricultural and biodiversity in certain progressive terms, like degraded lands restoration (Josephs et al. 2008). Generally, wild biodiversity is vulnerable by damage to habitat as soon as the area of crop production is extended. Though agronomic biodiversity is susceptible in instance of extensive mono-cropping, that is grounded on tapered genomic material pool also it prime to abridged usage of conventional diversities. The path because of biodiversity breach is meeting place deprivation succeeding land adaptation for feedstock production. Tropical regions are ideal for many biofuel crops (Curran et al. 2004; Soyka et al. 2007).

Crop genetic uniformity is the second major pathway for the loss of agrobiodiversity that is instigated by intensification on croplands. Mostly single species cultivation and plantations are possessed in biofuel feedstocks. Apprehensions are there about squat genetic diversity concerns are also found in sugarcane like grasses feedstocks (Luque et al. 2008), it causes the predisposition of crops to diseases and pests. Contrariwise, the converse is correct for the crops that retain an exceptional genetic diversity, much of that is unchanged, consequential a wide assortment of inherited physiognomies occur that dent its viable significance. If the second-generation feedstocks are genetically engineered to upsurge their efficiency, then they required a careful management within closed industrial production systems, or they need a lot of the enzymes for their conversion (Clancy 2008).

Constructive outcomes regarding biodiversity bear been stated durability, in debased or peripheral areas where instant enduring blended species have been possessed including reestablish biological system assignment then increase biodiversity. low-input, high-decent variability blends of local prairie perennials which propose a possibility of biotic communal managements, comprising wildlife terrain, carbon sequestration and water filtration equally supply greater net vivacity upsurges extra remarkable ozone diminishing element production reductions and a lesser amount of agrichemical contagion than do soybean-biodiesel or maize ethanol and that accomplishment escalations with the number of species. Switchgrass could be overpoweringly cost-effective on rich soils, predominantly when fertilizer and insecticides are associated, nonetheless that the situation accomplishment on the other side does not harmonize that of diverse confined perennials. In the marginal areas, the progressive effects on biodiversity have been noted, at these places, perennial mixed species are familiarized to upsurge the biodiversity and to refurbish the ecosystem working. Low contribution, high varietal mixture of intrinsic prairie perennials that propose a large number of ecosystem amenities, comprising wildlife habitat, carbon sequestration, and water filtration also generate high net energy gains larger GHG emission declines and low agrichemical pollution (Tilman et al. 2006).

10.5.4 Land Use Aspects

In recent years, utmost rise in inclusive agronomic products has resulted in 80% increase. The possible release effects of extending biofuel production will be met through enhanced land efficiency (Bruinsma 2003; Hazell and Wood 2008). Sustainable production of biofuel feedstocks with the advantage of use of barren land and abate the swift land-use change in areas previously under cultivation is possible by significant yield gains and technological advancements. Expansion of crop production in areas that are reserved for water filtration, biodiversity preservation, and carbon sequestration like important environmental services might result in harmful effects on environment. Available land for expanded crop production could be estimated after excluding the land and areas required to fulfill the amplified demand for livestock and food crops and forest land (Fischer et al. 2008).

Much land may be used forthrightly for the production of biofuel feedstocks; nevertheless extensive prevailing cropland could similarly elicit addition in the generation of non-biofuel crops at any other place. Previously uncultured land may wind up remarkably economical as ware costs increase, and the sparingly viable area would be needed to alter with extended attention toward new sources. Through management practices and advance technologies, the intensification of land use will have to accompaniment this opportunity, particularly if production is to be continuous (Bonifas et al. 2005; Pollock et al. 2008).

Biofuel yielding crops that endure ecological situations at places where food crops are not successful may propose the prospect to put the productive land practice that currently yields erratic commercial profits. Although marginal lands would be reduced prolific and subjected to high menaces, consuming them for bioenergy farms could have auxiliary benefits. It has been confirmed via various investigations that the economic yield of bioethanol from a good agricultural land is high than any additional costs. This creates the possibility of increasing interest and investment toward good land use that would be preceded in high payback acknowledge (Azar et al. 2006). Many crops are used to achieve economical feasible biofuel with the requirement of large water, fertilizers, and pesticides (FAO 2008).

10.6 Conclusion

To find non-conventional alternate resources of energy and to meet the energy demands various challenges need to be stated. Ethanol is biodegradable and less toxic to environment and it rapidly breaks down into innocuous affluences if spilled. Biomass ethanol feedstocks are able to improve the efficiency with which forestry and agricultural land are used in developing countries. If the accessible feedstock assets are used sustainably and implementable strategies are drawn in apt direction, it has a significant potential for biofuel production. Bioethanol absolutely has potential to reduce fossil resource use and greenhouse gas emanations, if resources like sugarcane and switchgrass are used.

References

Allen MR, Barros VR, Broome J, Cramer W, Christ R, Church JA, Clarke L, Dahe Q, Dasgupta P, Dubash NK (2014) IPCC fifth assessment synthesis report-climate change 2014 synthesis report

Ammann K (2013) GM crop risk debate, science and socioeconomics. Sustainable Food Production. Springer, pp 913–971

Arshad M, Hussain T, Iqbal M, Abbas M (2017) Enhanced ethanol production at commercial scale from molasses using high gravity technology by mutant *S. cerevisiae*. Braz J Microbiol 48(3):403–409

Azar C, Lindgren K, Larson E, Möllersten K (2006) Carbon capture and storage from fossil fuels and biomass–costs and potential role in stabilizing the atmosphere. Clim Change 74(1–3):47–79

Azizov AS (2006) The obtaining of fuel bioethanol. Appl Solar Energy 42(3):55

Baxter J, Bian Z, Chen G, Danielson D, Dresselhaus MS, Fedorov AG, Fisher TS, Jones CW, Maginn E, Kortshagen U (2009) Nanoscale design to enable the revolution in renewable energy. Energy Environ Sci 2(6):559–588

Bonifas KD, Walters DT, Cassman KG, Lindquist JL (2005) Nitrogen supply affects root: shoot ratio in corn and velvetleaf (*Abutilon theophrasti*). Weed Sci 53(5):670–675

Bruinsma J (2003) World agriculture: towards 2015/2030: an FAO perspective, Earthscan (ed.)

Chemtech S (2004) Pervaporation and vapor permeation technology. Application of pervaporation (brochure). Sulzer Chemtech, Switzerland
Cherubini F, Bird ND, Cowie A, Jungmeier G, Schlamadinger B, Woess-Gallasch S (2009) Energy-and greenhouse gas-based LCA of biofuel and bioenergy systems: key issues, ranges and recommendations. Resour Conserv Recycl 53(8):434–447
Cherubini F, Jungmeier G (2010) LCA of a biorefinery concept producing bioethanol, bioenergy, and chemicals from switchgrass. Int J Life Cycle Assess 15(1):53–66
Chum HL, Warner E, Seabra JE, Macedo IC (2014) A comparison of commercial ethanol production systems from Brazilian sugarcane and US corn. Biofuels Bioprod Biorefin 8(2):205–223
Clancy JS (2008) Are biofuels pro-poor? Assessing the evidence. Eur J Dev Res 20(3):416–431
Creutzig F, Ravindranath NH, Berndes G, Bolwig S, Bright R, Cherubini F, Chum H, Corbera E, Delucchi M, Faaij A (2015) Bioenergy and climate change mitigation: an assessment. Gcb Bioenergy 7(5):916–944
Curran LM, Trigg SN, McDonald AK, Astiani D, Hardiono Y, Siregar P, Caniago I, Kasischke E (2004) Lowland forest loss in protected areas of Indonesian Borneo. Science 303(5660):1000–1003
De Fraiture C, Giordano M, Liao Y (2008) Biofuels and implications for agricultural water use: blue impacts of green energy. Water Policy 10(S1):67–81
Demirbas A (2007) Progress and recent trends in biofuels. Prog Energy Combust Sci 33(1):1–18
Dixon B, Walker S, Golabi MH, Manner H (2012) Two probable latte period agricultural sites in northern Guam: their plants, soils, and interpretations. Micronesica 42(1/2):209–257
Doornbosch R, Steenblik R (2008) Biofuels: is the cure worse than the disease? Revista Virtual REDESMA 2:63
Dufey A (2006) Biofuels production, trade and sustainable development: emerging issues. ed. Iied
Enkvist P, Nauclér T, Rosander J (2007) A cost curve for greenhouse gas reduction. McKinsey Q 1:34
FAO B (2008) Prospects, risks, and opportunities. The state of food and agriculture [Online]. Food and Agrıculture Organızatıon of the Unıted Natıons, Rome
Fargione J, Hill J, Tilman D, Polasky S, Hawthorne P (2008) Land clearing and the biofuel carbon debt. Science 319(5867):1235–1238
Fischer M, Rudmann-Maurer K, Weyand A, Stöcklin J (2008) Agricultural land use and biodiversity in the Alps: how cultural tradition and socioeconomically motivated changes are shaping grassland biodiversity in the Swiss Alps. Mount Res Dev 28(2):148–155
Goswami M, Meena S, Navatha S, Rani KP, Pandey A, Sukumaran RK, Prasad R, Devi BP (2015) Hydrolysis of biomass using a reusable solid carbon acid catalyst and fermentation of the catalytic hydrolysate to ethanol. Biores Technol 188:99–102
Griggs DJ, Noguer M (2002) Climate change 2001: the scientific basis. Contribution of working group I to the third assessment report of the intergovernmental panel on climate change. Weather 57(8):267–269
Hazell P, Wood S (2008) Drivers of change in global agriculture. Philos Trans Royal Soc Lon B Biol Sci 363(1491):495–515
Josephs KA, Whitwell JL, Dickson DW, Boeve BF, Knopman DS, Petersen RC, Parisi JE, Jack CR (2008) Voxel-based morphometry in autopsy proven PSP and CBD. Neurobiol Aging 29(2):280–289
Kaminski W, Marszalek J (2006) Pervaporation for drying and dewatering. Drying Technol 24(7):835–847
Karthikeyan M, Zhonghua W, Mujumdar AS (2009) Low-rank coal drying technologies—current status and new developments. Drying Technol 27(3):403–415
Kelly Y, Tau L, Lee KT, Mohamed AR, Bhatia S (2007) Potential of hydrogen from oil palm biomass as a source of renewable energy worldwide. Energy Policy 35(11):5692–5701
Kim S, Dale BE (2002) Allocation procedure in ethanol production system from corn grain I. system expansion. Int J Life Cycle Assess 7(4):237–243

Kim S, Dale BE (2004) Global potential bioethanol production from wasted crops and crop residues. Biomass Bioenergy 26(4):361–375
Koh LP, Ghazoul J (2008) Biofuels, biodiversity, and people: understanding the conflicts and finding opportunities. Biol Cons 141(10):2450–2460
Kumar A, Singh D, Chandel AK, Sharma KK (2017) Technological advancements in sustainable production of second generation ethanol development: an appraisal and future directions. Sustainable biofuels development in India. Springer, pp 299–336
Li X-K, Fang Z, Luo J, Su T-C (2016) Coproduction of furfural and easily hydrolyzable residue from sugar cane bagasse in the MTHF/aqueous biphasic system: influence of acid species, NaCl addition, and MTHF. ACS Sustain Chem Eng 4(10):5804–5813
Lin C-Y, Lay C-H, Sen B, Chu C-Y, Kumar G, Chen C-C, Chang J-S (2012) Fermentative hydrogen production from wastewaters: a review and prognosis. Int J Hydrogen Energy 37(20):15632–15642
Liu Z, Zhang C, Lu Y, Wu X, Wang L, Wang L, Han B, Xing X-H (2013) States and challenges for high-value biohythane production from waste biomass by dark fermentation technology. Biores Technol 135:292–303
Luque R, Herrero-Davila L, Campelo JM, Clark JH, Hidalgo JM, Luna D, Marinas JM, Romero AA (2008) Biofuels: a technological perspective. Energy Environ Sci 1(5):542–564
Marszałek J, Kamiński W (2008) Environmental impact of bioethanol production. Proc Ecopole 2(1):65–70
Marszałek J, Kamiński W, Środowisko BN (2009) Environmental impact of bioethanol production. Ecol Chem Eng Sci 16(4):429–438
Mathur D, Medintz IL (2017) Analyzing DNA nanotechnology: a call to arms for the analytical chemistry community. ACS Publications
Mirza J, Riaz M, Naseer A, Rehman F, Khan A, Ali Q (2009) Pakistani bentonite in mortars and concrete as low cost construction material. Appl Clay Sci 45(4):220–226
Ohara H (2003) Biorefinery. Appl Microbiol Biotechnol 62(5–6):474–477
Pate R, Klise G, Wu B (2011) Resource demand implications for US algae biofuels production scale-up. Appl Energy 88(10):3377–3388
Pollock C, Pretty J, Crute I, Leaver C, Dalton H (2008) Introduction. Sustainable agriculture. The Royal Society
Rhodes CJ (2014) Soil erosion, climate change and global food security: challenges and strategies. Sci Prog 97(2):97–153
Righelato R, Spracklen DV (2007) Carbon mitigation by biofuels or by saving and restoring forests? Sci N Y Washing 317(5840):902
Romano R (1998) Empowering investors: a market approach to securities regulation. Yale Law J 107(8):2359–2430
Ryan L, Convery F, Ferreira S (2006) Stimulating the use of biofuels in the European Union: implications for climate change policy. Energy Policy 34(17):3184–3194
Sander U, Soukup P (1988) Design and operation of a pervaporation plant for ethanol dehydration. J Membr Sci 36:463–475
Sander U, Soukup P (1991) Practical experience with pervaporation systems for liquid and vapour separation. J Membr Sci 62:67–89
Shah SH, Raja IA, Rizwan M, Rashid N, Mahmood Q, Shah FA, Pervez A (2018) Potential of microalgal biodiesel production and its sustainability perspectives in Pakistan. Renew Sustain Energy Rev 81:76–92
Shapouri H, Duffield J, Wang M (2003) The energy balance of corn ethanol revisited. Trans ASAE 46(4):959
Silalertruksa T, Pongpat P, Gheewala SH (2017) Life cycle assessment for enhancing environmental sustainability of sugarcane biorefinery in Thailand. J Clean Prod 140:906–913
Silveira MHL, Morais ARC, da Costa Lopes AM, Olekszyszen DN, Bogel-Łukasik R, Andreaus J, Pereira Ramos L (2015) Current pretreatment technologies for the development of cellulosic ethanol and biorefineries. Chemsuschem 8(20):3366–3390

Sitnikov CS (2013) Corporate strategy. Encyclopedia of corporate social responsibility, pp 606–613
Solomon BD, Birchler J, Goldman SL, Zhang Q (2014) Basic information on maize. In: Corn SLG, Kole C (eds) Compendium of bioenergy plants. CRC Press, Boca Raton, FL, pp 1–32
Soyka T, Palmer C, Engel S (2007) The impacts of tropical biofuel production on land-use: the case of Indonesia. In: Conference on International Agricultural Research for Development
Suh K, Suh S (2010) Economic and environmental implications of corn stover densification options for biofuel in Minnesota. Trans ASABE 53(4):1183–1192
Sun Y, Cheng J (2002) Hydrolysis of lignocellulosic materials for ethanol production: a review. Biores Technol 83(1):1–11
Tilman D, Hill J, Lehman C (2006) Carbon-negative biofuels from low-input high-diversity grassland biomass. Science 314(5805):1598–1600
Wang M, Wu M, Huo H (2007) Life-cycle energy and greenhouse gas emission impacts of different corn ethanol plant types. Environ Res Lett 2(2):024001
Wilhelm WW, Johnson JM, Karlen DL, Lightle DT (2007) Corn stover to sustain soil organic carbon further constrains biomass supply. Agron J 99(6):1665–1667
Woodward A, Smith KR, Campbell-Lendrum D, Chadee DD, Honda Y, Liu Q, Olwoch J, Revich B, Sauerborn R, Chafe Z (2014) Climate change and health: on the latest IPCC report. The Lancet 383(9924):1185–1189
Yeh S, Farrell A, Plevin R, Sanstad A, Weyant J (2008) Optimizing US mitigation strategies for the light-duty transportation sector: what we learn from a bottom-up model. ACS Publications
Zhang K, Pei Z, Wang D (2016) Organic solvent pretreatment of lignocellulosic biomass for biofuels and biochemicals: a review. Biores Technol 199:21–33

Printed in the United States
by Baker & Taylor Publisher Services